Retrofitting of Heritage Structures

Design and evaluation of strengthening techniques

Retrofitting of Heritage Structures

Design and evaluation of strengthening techniques

Edited by

S. Syngellakis
Wessex Institute of Technology, UK

WIT*PRESS* **Southampton, Boston**

Editor:

S. Syngellakis
Wessex Institute of Technology, UK

Published by

WIT Press
Ashurst Lodge, Ashurst, Southampton, SO40 7AA, UK
Tel: 44 (0) 238 029 3223; Fax: 44 (0) 238 029 2853
E-Mail: witpress@witpress.com
http://www.witpress.com

For USA, Canada and Mexico

WIT Press
25 Bridge Street, Billerica, MA 01821, USA
Tel: 978 667 5841; Fax: 978 667 7582
E-Mail: infousa@witpress.com
http://www.witpress.com

British Library Cataloguing-in-Publication Data
A Catalogue record for this book is available
from the British Library

ISBN: 978-1-84564-754-4
eISBN: 978-1-84564-755-1

Library of Congress Catalog Card Number: 2012953674

Printed by Lightning Source, UK.

Preface

The preservation of heritage architecture is a cultural objective rigorously pursued by communities and nations wishing to promote their history, civilisation and aesthetic achievements. Structures built in the remote past by traditional methods have suffered the consequences of extreme loading events, such as earthquakes, over long time periods. Retrofitting is an approach based on recent technological developments and scientific knowledge, whereby modern construction methods and materials are applied to the repair and strengthening of historical structures. This book, based on an edited collection of articles from the Transactions of the Wessex Institute, aims to inform on current retrofitting techniques, their application to various types of historical architecture and their effectiveness to fulfil their purpose.

The adoption of particular retrofitting techniques depends on the type and size of the structure to be preserved as well as the conditions it is protected from. Structural forms vary widely from age old places of worship, such as churches, mosques and temples, as well as castles and palaces to more modern, distinguished private residences or public buildings, some of them designed by well known architects. The methods of constructions range from traditional, such as stone or brick masonry to more recent textile block systems and reinforced concrete frameworks.

A detailed visual inspection of a damaged structure provides a valuable insight into possible causes of failure; these are often attributed not only to inadequate strength of the original construction but also to subsequent not well thought attempts to strengthen it. Testing samples extracted from the existing structure leads to material characterisation which is essential to modelling for performance assessment. The behaviour of the structure before and after retrofitting can be assessed by numerical analysis usually based on general purpose finite element packages. The analysis of the original structure informs on its existing strength and weaknesses and assists the prospective retrofitting plan and design. The analysis of a structure strengthened by either conceptual or actual retrofitting serves as an evaluation of the effectiveness of the proposed or taken measures.

Strengthening techniques for masonry walls include re-pointing, injection grouting and the use of steel ties. Reinforced concrete is used in the form of cast-in-place walls, jackets around column or beams or tie-beams at floor and foundation levels; carbon fiber reinforced laminates can be used for reinforcing walls and slabs. Innovative use of materials, such as shape memory alloys, self-compacting concrete or thin lead layers is also suggested.

Earthquakes arguably constitute the most damaging external action and most of the research reported in the book focuses on methods to moderate their consequences. Seismic energy absorbing devices and base isolation systems are two effective means of providing protection against future seismic events although their application is often met with many technical challenges in practice. This book addresses, to a certain extent, all the above issues; it is thus expected to make an important and topical contribution to the body of knowledge in this area.

Stavros Syngellakis
The New Forest, 2013

Contents

Effectiveness of seismic retrofit for stone masonry structures

V. Kilar[1] & D. Marušić[2]
[1]*Faculty of Architecture, University of Ljubljana, Slovenia*
[2]*Dinamika–Ideja–Prostor Ltd. (Dipstor), Ljubljana, Slovenia*

Abstract

This paper analyses the effectiveness of various seismic retrofitting techniques for the renovation of stone masonry structures. The first part of the paper summarizes the most widely used principles for strengthening buildings made of natural stone units or a mixture of mortar and natural stone aggregate. In the second part of the paper some selected retrofitting techniques are applied for the structural and seismic renovation of a historical object (church built around 1570 located in Solčava in Slovenia). Before 2006, the church was critically cracked and not suitable for use. The retrofitting study carried out by the authors included the mandatory and recommended measures that should have taken place immediately to prevent the settling down of the foundations and increase the seismic resistance of the church. To date only partial strengthening of the foundations has taken place. The behavior of the church at various renewal stages was verified with a numerical analysis of a very refined 3D computer model of the church. Three models were analyzed: (a) a model of the existing church including the settling down of the SE wall, (b) a model of the existing church after the partial retrofitting of the foundations in 2006 and (c) a model as (b), including the minimum suggested retrofitting proposals.
Keywords: stone masonry structures, structural retrofit, seismic retrofit, seismic resistance, historical object, church building.

1 Overview of the methods for seismic retrofit of historical stone masonry structures

The cracks, subsiding of foundations and inadequate horizontal stiffness are among the most common causes that call for structural/seismic retrofit of historical buildings. The seismic strengthening of a building requires not only the increase of stiffness, but also the increase of ductility of the individual elements as well as the structure as a whole. The increase of stiffness is generally related to the increase of earthquake forces and can produce an unfavorable result if the ductility is not increased along with the stiffness. Special attention should be given to partial reconstruction measures where the increase of stiffness of one element might cause the unfavorable redistribution of internal forces during an earthquake. The retrofitting proposals should therefore always be supported with adequate numerical analyses, which prove the effectiveness of proposed measures. Some most widely used retrofitting measures for stone masonry objects are briefly summarized.

Retrofitting of the foundations

- Consolidation of the foundations with the injection of cement-silicate mixture (might include also hydrophobic additives).
- Widening of the existing foundations.
- Injecting of concrete under foundations and filling the voids under them.
- Injecting of expansion materials and lifting of the foundations.
- Solutions with pile systems, etc.

Strengthening of stone masonry walls

- Injecting of cement-silicate mixture into the hollow middle part of the wall with moderate pressure (up to 2 bar).
- Concreting of RC concrete plaster enclosed and anchored to old wall.
- Adding of the vertical RC confining elements.
- Adding of vertical steel prestressed cables.
- Strengthening of walls with plastic ribbons or textile bands made of GFRP (Glass Fiber Reinforced Polymer) or more expensive CFRP (Carbon Fiber Reinforced Polymer).
- Some development has been made also in the field of seismic isolation with seismic bearings and dampers which activate only during stronger earthquakes.

Interconnection of walls

- The interconnection of walls with horizontal steel bars anchored to steel plates.
- Adding of horizontal confining RC elements.
- Insertion of corner triangular stone elements and injecting of walls.
- Adding of steel ribbons and injecting.

Retrofitting of arches

- Injecting of cracks.
- Rebuilding of damaged parts.
- Removing of heavy filling and fabricating of RC plate.
- Filling with reinforced aero-concrete.
- Strengthening of arches with composite polymer ribbons.

2 Analyzed stone masonry church

2.1 Description of church in Solčava

The church of "Marija Snežna" was built in the second half of the 15th century (fig. 1, left). In past centuries it was repaired several times and according to our information, the last time it was repaired thoroughly was in 1824. The arch thickness is constant and it amounts to 37 cm. The thickness of walls varies from 72 up to 120 cm. Today the church is heavily cracked; wide cracks of walls can be seen on the outer and inner sides as well as cracks in arches and stone groins (fig. 1, middle). The main reason is the subsiding of the NE wall of the main nave and presbytery (side opposite to bell tower), which has weaker foundations than other parts of the church. As a primary rescue measure, the strengthening of the foundations was first proposed in order to prevent the threatened vault collapse (partly completed at the end of 2006). Other recommended retrofit measures are described and discussed in the last part of the paper.

Figure 1: Church in Solčava (Slovenia): photos and mathematical model.

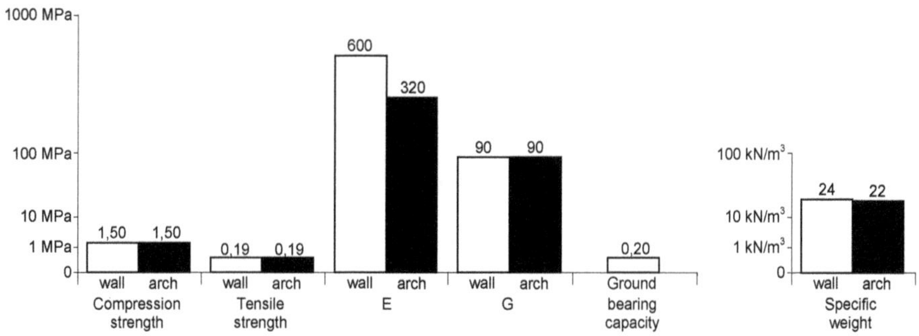

Figure 2: Stone masonry and ground characteristics (cube root scale).

2.2 Material and ground characteristics

The church is built of stone aggregate connected with lime mortar. The size of the stone pieces in walls varies from small pieces with maximum diameter of up to 25 mm to larger stone units reinforcing the corners and vertical strengthening columns. The arches are built of stone grains with maximum diameter of 20 mm. According to the results of material testing performed by IRMA in 2004, the compression strength of lime mortar varies from 1.0 – 1.5 MPa for the walls and from 1.5 – 2.0 MPa for the vaults and tower [1].

Based on these data and obtained moisture gravimetrical measurements (ranging from 6.2 to 10.6%), the compression and tension strengths and elastic and shear modulus of walls and arches have been estimated. These values were further compared with the recommendations given by Tomaževič [2]. In the present analysis, the average values obtained from both sources have been used. The used material characteristics are summarized in fig. 2. The allowable ground stress obtained from the geotechnical report [3] amounts to 200 kPa.

2.3 Mathematical model

The church was modeled with the computer program SAP2000 [4]. The maximum side size of 2D planar 4-node and 3-node elements was limited to 40×40 cm. The modeling of groins and arches connections was particularly complicated and required adequate refinement of the numerical model mesh. The number of all used finite elements was around 21,000. For practical reasons the arches were modeled separately from the walls. The arch reactions were transferred to the joints at the tops of the walls. The whole wall model with the bell tower was then loaded also by the reactions transmitted by the wooden roof construction. The existing steel ties were also considered separately. It was assumed that the arches are pinned to the walls and that the walls are fixed onto the foundations. The arch photos and their mathematical model are presented in fig. 3.

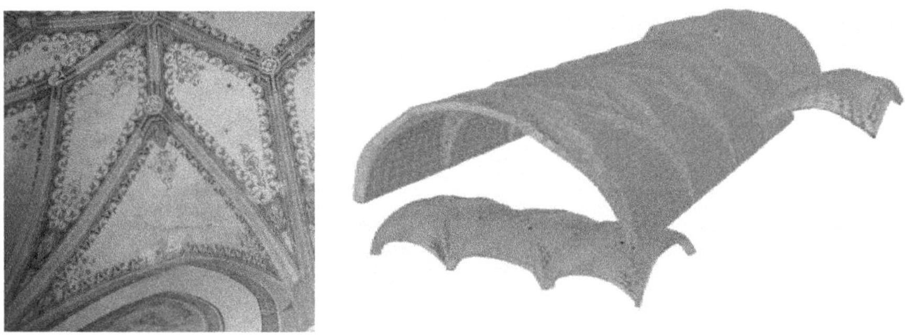

Figure 3: Arches: photo and mathematical model.

2.4 Vibration modes

The numerical model should include all vibration modes that significantly influence the response. According to Eurocode 8 [5], this requirement is fulfilled if all modes associated with an effective mass of at least 90% of the whole mass or all modes with effective mass greater than 5% of the whole mass are taken into account. In this case the influence of higher modes is extremely important; 24 vibration modes had to be considered. The first three modes are presented in fig. 4. The first two fundamental modes of the bell tower amount to 1.38 s

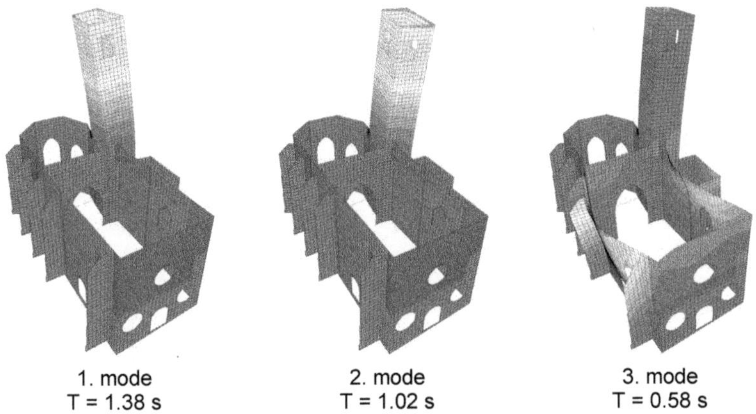

Figure 4: First three fundamental modes.

(direction N-S) and 1.02 s (direction E-W). The movement of the top of walls becomes significant only at the third fundamental mode (0.58 s) or at higher modes.

2.5 Loading

The church is subjected to dead load, live roof load and earthquake loading in the longitudinal *X* (along the main nave) and transversal *Y* direction. The EC8 response spectrum for type B soil class with peak ground acceleration 0.15 g was used. The selected results of the analyses present the envelope of vertical loading and $\pm X$, $\pm Y$ loads along with ±SRSS combination of earthquake loading in *X* and *Y* direction.

3 Analysis of existing church subjected to vertical displacement of NE wall

The existing church is heavily cracked due to the subsiding of the wall foundations on the NE side of the main nave (opposite to the bell tower). The selected analysis results of the existing church subjected to a vertical displacement of the foundations of the NE wall foundation of 2 cm (actual displacement obtained at the spot) are presented in fig. 5. For the arches the maximum main stress combination at the top side of the arches is shown. For the walls the stress σ_{11} acting in local direction 1 (for all elements it lies in horizontal plane) and the stress σ_{22} acting in local direction 2 (for all elements it lies in vertical plane) are shown. Thus, the stress σ_{11} might cause the cracks propaginating in the vertical direction and σ_{22} the cracks in the horizontal direction. Considered allowable stresses are σ_{dop1}=0.19 MPa in tension and σ_{dop2}=1.50 MPa in compression. The results show that the tensile allowable stress was exceeded in several parts of the main nave and presbytery, mainly at the spots where the cracks actually appeared. The subsiding of the NE wall would be critical also for the stability of the vaults. The church in its existing state is practically statically unstable. Every subsequent subsiding of foundations might cause the collapse of the church vaults. For this reason an immediate strengthening of the foundations was suggested; this took place at the end of 2006. The analysis of the church with retrofitted foundations is presented in section 4.

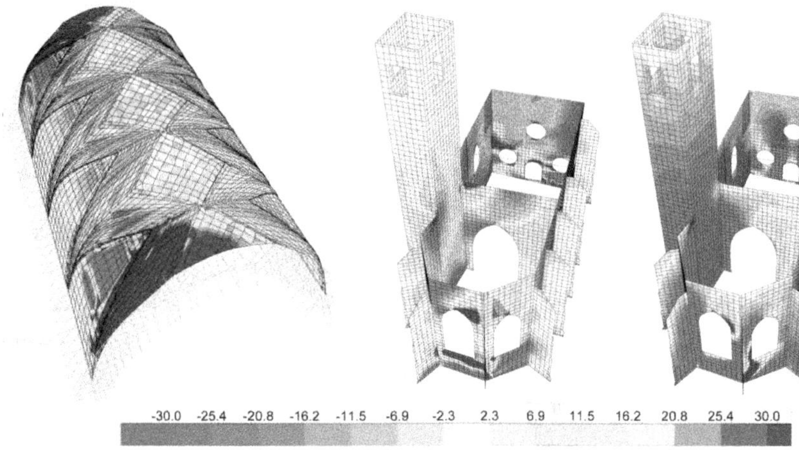

Comb. of stresses σ_{11} and σ_{22}. Stress σ_{11}. Stress σ_{22}.

Figure 5: Stresses in the arch of main nave and in church walls due to vertical displacement of NE wall for 2 cm (10-2 MPa).

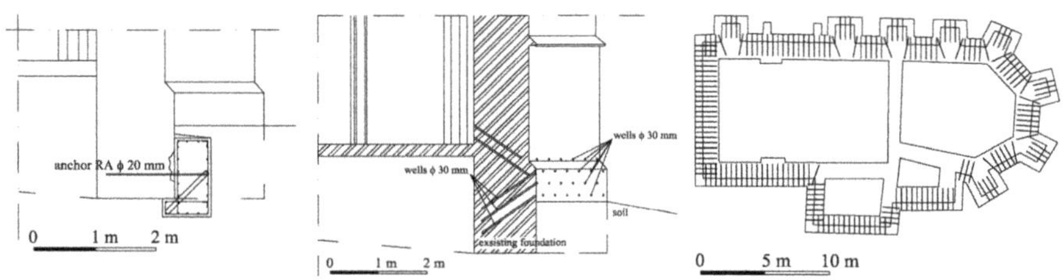

New and old foundations Injection holes Positions of anchors

Figure 6: Design project of the new foundations (LEGRO, 2006).

4 Analysis of existing church after the retrofitting of foundations

4.1 Retrofitting of the foundations in 2006

The actual vertical stress under the NE wall of the main nave of the church amounted to 270 kPa and exceeded the allowable stress (200 kPa) by approximately 30%. The eccentricity of the vertical resultant (25.5 cm) fell slightly out of the existing foundation core area (23.3 cm). The analysis results showed that the foundation should be of at least 2.4 m width. The retrofitting of the foundations took place at the end of 2006 (fig. 6 and 7). After the injection and widening of the existing foundations, the injection of the ground below them was also performed. The injection works were performed by Uretek© using their floor lifting method (<www.uretek@uretek.si>).

Figure 7: New foundations and injecting of lifting resin compound under the existing foundations (LEGRO and URETEK. Ltd).

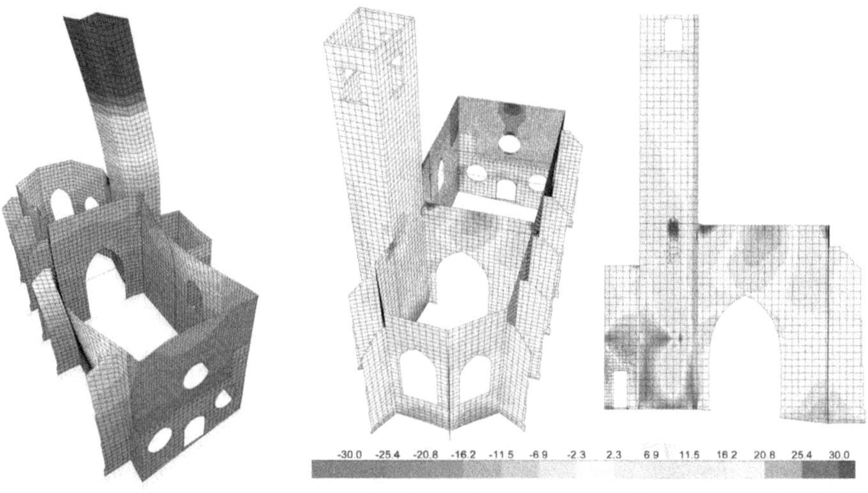

Displacement (max U_y=1.79 cm). Stress σ_{11}.

Figure 8: Maximal displacement of bell tower and stresses in church walls due to vertical and earthquake loading (10^{-2} MPa).

4.2 Analysis of the existing structure with retrofitted foundations

The analysis results for the vertical dead and live load show that the allowable stresses are not exceeded in the case of vertical loading. However, the combination with earthquake loading according to EC8 shows that the church is absolutely unsafe for eventual earthquake threat. The deformation shape for Y direction displacement and stress σ_{11} in the walls are presented in fig. 8. The maximum X and Y displacements of the tower top amount to U_x = 1.17 cm and U_y = 1.79 cm. The maximum displacement at the top of the wall, perpendicular to the wall plane, amounts to 0.63 cm (in the front entrance wall with the rosette). The allowable tensile stress σ_{11} is exceeded in a number of places, especially at the top of the walls and connections of bell tower and church.

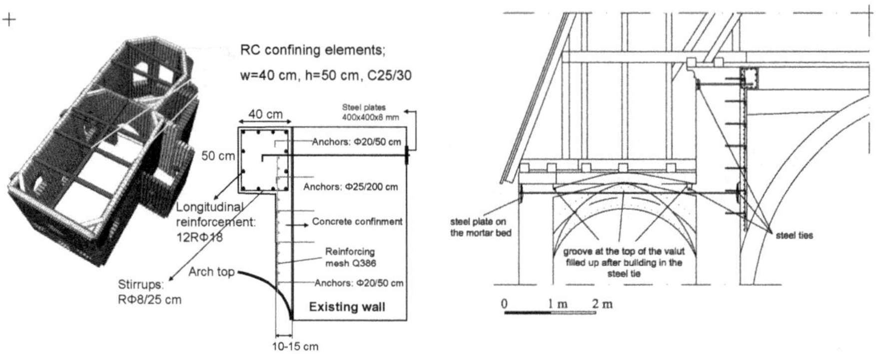

Figure 9: Required retrofitting construction measures (RC confining elements at the top of walls above the arches, steel anchors, wall concreting).

5 Analysis of retrofitted church

5.1 Proposed measures of seismic retrofit

5.1.1 Confining and transversal RC beams simulating a "rigid diaphragm"

According to the results obtained in section 4 it can be concluded that the most effective retrofit measure must be the connecting of tops of walls by a confinement system that simulates a "rigid diaphragm". In the main nave, the system of confining RC beams at the tops of walls with additional transversal cross beams above the arches has been suggested (fig. 9). The confining RC beams must be added around all wall tops in the space between the vault base and vault top. They must be anchored to the church walls with steel bars with diameter 20 mm and every 2 m with the screw anchors with bolts fixed to the steel plate on the other side as shown in fig. 9.

5.1.2 Injection of walls

The walls are heavily cracked and their load capacity can be increased by injection of cement mixture. This procedure is however problematic, because the church walls are covered with valuable gothic paintings on the inner and outer sides. The given preservation recommendations suggest the usage of injection compounds which do not contain the cement as well as a limitation of the maximum injection pressure.

5.1.3 Retrofitting of arches and groins

The main arches and groins/ribs will be stabilized by the proposed confining RC beams (section 5.1.1), which will accommodate all horizontal reactions of the arches. In this way the vaults, which primarily act in compression, will remain stable. One of the options for strengthening the arches is also the usage of carbon ribbons; however the ribbons do not adhere easily to stone masonry units. As another option, the fabrication of thin concrete shell above the existing arches might also be planned; this can be supported by the external walls. This proposal might be contradictory to already mentioned conservatory requirements.

5.1.4 Vertical confining elements

The seismic response of the church could be further improved by adding reinforced concrete vertical confining elements, which would increase the ductile wall behavior as well as enable a direct transfer of horizontal forces to the foundations. The vertical confining elements are however difficult to be built without the roof removal or without interventions to visual wall surface. One of the alternative options would be to build in the vertical steel rods (e.g. prestressed cables) and fix them to the steel plates at the wall tops and anchor them to the foundations.

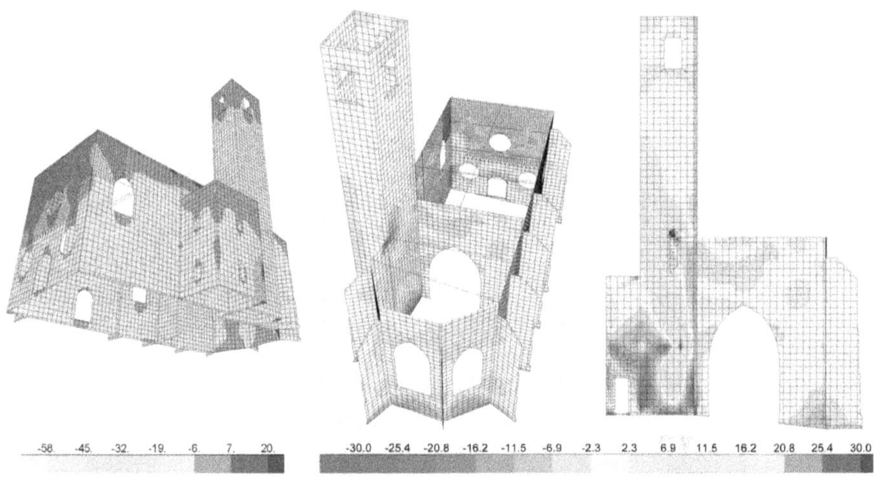

Stress σ_{22} Stress σ_{11}

Figure 10: Stresses in retrofitted church due to vertical and earthquake loading (10^{-2} MPa).

5.2 Analysis of the seismic retrofitted church

The analyzed model of the retrofitted church includes only the proposed RC confining elements, which might simulate a rigid horizontal diaphragm as described in section 5.1.1. Other proposed measures were not taken into account. The selected analysis results are presented in fig. 10. It can be seen that the

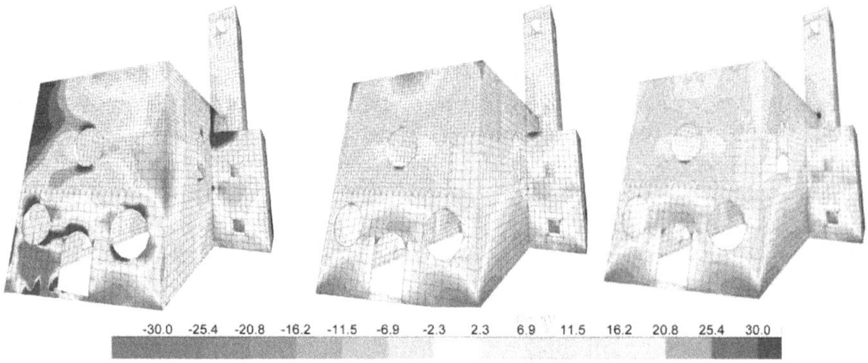

Old church with settlement of NE wall – potentially unstable – before 2006.

Existing situation from the end of 2006 – retrofitted foundations.

Renovated church structure according to authors' proposals.

Figure 11: Comparison of stress σ_{11} for different renovation stages (frontage: view from west side) due to self-weight and earthquake (10^{-2} MPa).

behavior of the church (except for a few spots at the junction of the tower and the wall) remains elastic during the design earthquake.

6 Conclusions

The numerical analyses covered three renewal stages of the church in Solčava, e.g. (a) original church with subsiding of the NE wall foundation, (b) church with new retrofitted foundations and (c) church with the proposed RC confining elements simulating a rigid diaphragm at the top of the walls (fig. 11). It was shown that the church in stage (a) is practically unstable and that the church in stage (b) is stable for vertical loading, but it is not capable of withstanding the earthquake loading according to EC8 (note that this is also the present state of the church at the beginning of 2007). Stage (c), which includes the RC horizontal confining elements according to the authors' proposal, proves to be adequately safe fulfilling the EC8 requirements. It can be concluded that the proposed measures should be carried out as soon as possible.

Acknowledgements

The authors would like to express their gratitude to the companies which were in charge of the preparation of detailed projects and realization of the works in Solčava, as well to the individuals who shared their experience as well as their technical and photo material. In the first instance, thanks are due to Ing. Ivan Golob (LEGRO kd), Ing. Franc Vidic (GEOTEC Ljubljana, Ltd.) and to companies IRMA (Institute for Research of Materials and Applications Ltd.), DIPSTOR, Ltd., Ljubljana and Faculty of Architecture in Ljubljana.

References

[1] Leskovar, I., Grčar, S. & Šuštešič, J., Poročilo o preiskavah kamnitih zidov na objektu farna cerkev Marije Snežne, Solčava. IRMA d.o.o. Ljubljana: Ljubljana, 2004 (Material investigation report).
[2] Tomaževič, M., Zidane zgradbe na potresnih območjih. *Univerza Edvarda Kardelja v Ljubljani*, Fakulteta za arhitekturo, gradbeništvo in geodezijo: Ljubljana, 1987.
[3] Vidic, F., Geotehnično poročilo o pregledu razpok, opredelitvi zemljin ter predlogu sanacije temeljenja cerkve Marije Snežne v Solčavi. *GEOTEC Ljubljana*, d.o.o., Ljubljana, 2004 (Geotechnical report).
[4] Computers and Structures, Inc. SAP 2000 Nonlinear, Version 10. Computers and Structures, Inc: Berkeley, California, USA, 2006.
[5] Eurocode 8. Design of structures for earthquake resistance Part 1: General rules, seismic actions and rules for buildings, European Standard EN 1998-1: 2004. CEN, European Committee for Standardization.

Retrofitting of the Soltani Mosque after the Silakhor Plain earthquake

H. R. Vosoughifar & S. K. Sadat Shokouhi
Islamic Azad University, South Tehran Branch, Tehran, Iran

Abstract

In this paper the retrofitting process of the Soltani (Imam) Mosque after the Silakhor Plain earthquake (31 March 2006) was evaluated. Retrofitting historical structures or monuments has been a challenge for experts and authorities for years while no basic and integrated measures are taken. The earthquake tremors in Boroujerd city caused damage of various degrees in the historic and cultural buildings. In the city of Boroujerd, the most significant damage included that of the minarets in the Jame' mosque, the collapse of false ornamental stalactite ceiling in the Soltani (Imam) mosque and partial collapse of the beehive dome in the Imamzadeh Ja'far shrine in the Lorestan province. The main damage to the Soltani Mosque is the collapse of the false ceiling including the rich stalactite ornament in the south *eivan* and cracking in the flanking parts of the north *eivan*. Destruction resulted from the addition of a big concrete beam over the doorway that changed the behaviour of the latter and made it rigid. The concrete tie-beam in the south *eivan* of the Imam mosque has aggravated the impact of an earthquake on the structure as, during construction, the keystone of the original arch has also been weakened. This tie-beam should be replaced by a more resilient system of reinforcement. Moreover, terms of reference for cooperation between restoration and retrofitting of the monuments should be drafted specifically for each monument according to the particular conditions prevailing in each case.
Keywords: seismic retrofitting, Soltani Mosque, monuments, earthquake.

1 Introduction

A new interest in masonry buildings, especially the historical ones, has arisen in the past years all over Europe and Asia as a consequence of the need for restoring ancient or historically interesting structures while preserving their main features and guaranteeing, at the same time, their strength. Besides, old masonry buildings form a significant part of the building stock in many cities and they are still used for either housing or services, hence they need a sufficient level of safety with respect to both the vertical loads and possible seismic events. Furthermore, some structural modifications, which may have been made through the centuries, often make difficult the evaluation of the actual structural vulnerability of either the whole complex or some parts of it [1].

Ancient masonry structures are particularly vulnerable to dynamic actions, with a special focus on seismic action. Due to the ageing process as well as to the environmental factors, many cultural heritage buildings, as structures planned and constructed in the past, result to be vulnerable to dynamic loads, which may unpredictably induce a collapse of a portion or drive the whole structure to a rapid failure. But the high vulnerability of historical masonry buildings to seismic actions is mostly due to the absence of adequate connections between the various parts (masonry walls, timber beams in the floors and timber beams in the roof). This characteristic leads to overturning collapse of the perimeter walls under seismic horizontal acceleration [2].

The assessment of the seismic capacity of historic masonry buildings presents objective difficulties deriving from the analytical treatment of the masonry material non-linearity, which displays nearly no-tension characteristics and requires proper experimental data for the calibration of the numerical models. In addition, the complexity of geometrical configuration of this building typology often requires the implementation of models characterized by a large number of degrees of freedom. Therefore a full non-linear dynamic analysis of masonry buildings is not an immediate task. Retrofitting existing structures to resist seismic actions that they were not originally designed for is a common practice in structural engineering. Recently, recommendations for the analysis, conservation and structural restoration of architectural heritage have been approved by ICOMOS [3]. These recommendations are intended to be useful to all those involved in conservation and restoration problems and not exclusively to the wide community of engineers. A key message, probably subliminal, is that those involved in historic preservation must recognize the contribution of the engineer. Often engineering advice seems to be regarded as something to be sought at the end of a project when all the decisions have been made, while it is clear that better solutions might have been available with an earlier engineering contribution.

Modern intervention procedures require a thorough survey of the structure and an understanding of its history. Any heritage structure is the result of the original design and construction, any deliberate changes that have been made and the ravages of time and chance. Structures of architectural heritage, by their very nature and history (material and assembly), present a number of challenges in conservation, diagnosis, analysis, monitoring and strengthening that limit the application of modern legal codes and building standards [4].

During the past decade, many valuable heritage buildings have been damaged or destroyed by earthquakes in Iran, notably in Bam [5], Tabas and recently in

Boroujerd. On 31 March 2006, a series of earthquakes with the strongest shock measuring 6 on the Richter scale (according to Iran Geophysics Centre) struck south-western Iran and affected the cities and villages around Boroujerd city of Lorestan province. The seismic jolt caused extensive damage in many villages and the city of Boroujerd. There are about 40 cultural heritage properties which have sustained damages of various degrees in the earthquake-stricken area.

The knowledge of the repair and conservation of historical buildings is old, and sufficient experience exists in this area. The earthquake epicentre was in Darb-e-Astaneh, a remote village about 40 km west of the city of Doroud. The series of seismic shocks however affected a vast populated rural and urban area: some 25 villages and the city of Boroujerd were severely damaged. Telephone lines, electricity and gas supplies had been cut in some areas. The city of Boroujerd dates to ancient history of Iran. Containing approximately 40 cultural heritage properties, it boasts most of the built cultural property in the whole Lorestan province.

2 Historic earthquakes in the region

Iran is located on major geological fault lines along Alborz and Zagros mountains and is regularly struck by powerful earthquakes. Zagros Mountains include the Zagros fault which plays a significant role in the seismicity of Iran. The direction of the Zagros fault is extended from the Turkey border to east Haji Abad of Bandar Abbas, north western-south eastern (N130E) while it is turning in this area. From this point to the south, the Zagros fault with a length of 1350 km follows a north western-south eastern direction (N170E). Young faults which coincide with the main Zagros fault have been called Main Recent Fault (MRF), which is coincident with the old fault. This fault is not a single structure but it is a narrow zone of individual, separated and generally right turning fault parts with an echelon design. The current main fault parts from south east to North West are: Doroud fault, Nahavand fault, Garon (Karon) fault, Sahneh fault, Morvarid fault and Piranshahr fault. The main Zagros fault and the Main Recent Fault have been detected as an end sliding, right turning and seismic generating fault with a north east- south west direction after the northeast line of Main Zagros Reverse Fault. Geological evidences support the right turning displacement up to 10-60 km by Doroud and Nahavand parts of the current fault. It is worth noting that parts of the main young fault which led to right turning movement of reverse main Zagros fault (Doroud, Nahavand, Sahneh and Dinor parts of MRF) have more seismic activity than other parts of this fault (Sartakht, Morvarid, Marivan and Piranshahr parts). The 31st March earthquake occurred close to the Main Recent Fault, a major strike-slip fault, but it is not known at present if the earthquake occurred on that fault or on a related smaller fault. This fault along its length from Dinor in north-eastern, has been interrupted with the geological event of Farsinag seismic to Gahar lake in north-eastern with Silakhor earthquake (1/23/1909, Ms=7.4) and historical earthquakes in the area. The section of the main fault near the epicentre of the 31st March earthquake produced a magnitude 7.4 earthquake in 1909 destroying the village of Silakhor plain.

The first tremors, with a magnitude of 4.7 at 7:45 pm on 30/03/06 were followed by one of 5.1 at 01.05 am on Friday (21:35 GMT on Thursday) and by another of 6.1 at 04.47am, the official Iran news agency reported. The International Institute of Earthquake Engineering and Seismology (IIEES) identifies Darb-e-Astaneh (Silakhor) region, located between Boroujerd and Doroud cities, as the epicentre and indicates that the earthquake occurred on a smaller fault related and close to the main fault. Fig. 1 illustrates detailed location maps of faults showing the position of the fault in relation to Doroud.

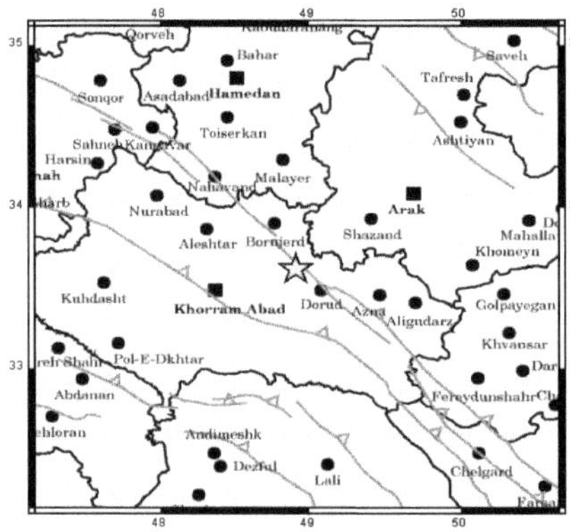

Figure 1: Detailed location maps of faults showing position of the fault in relation to Doroud.

3 Silakhor plain earthquake and monuments

On 31 March 2006, a series of earthquakes with the strongest shock measuring 6 on the Richter scale struck south-western Iran and affected the cities and villages around Boroujerd city in the Lorestan Province. There are about 40 cultural heritage properties which have sustained damage of various degrees in the earthquake-stricken area. The earthquake epicentre was in Darb-e-Astaneh a remote village about 40 km west of the city of Doroud. The series of seismic shocks however affected a vast populated rural and urban area.

In the city of Boroujerd, the most significant damages include the damage of the minarets in the Jame' Mosque, the collapse of false ornamental stalactite ceiling in the Soltani (Imam) Mosque and partial collapse of the beehive dome in the Imamzadeh Ja'far Shrine. In the Lorestan province, Imamzadeh Khaled Ibn Ali, Hojatieh School, Ghaleh and Rangineh Mosques were also damaged. Many of the old historic houses were also damaged in the historic fabric of Boroujerd city. However, rapid urbanization and building activities during the past three decades have caused soaring land prices and created the tendency to convert the cultural and historic property into modern apartment buildings accessed by wide streets. Lack of vehicular access in the old fabric, lack of proper maintenance, misuse or inappropriate upkeep and use, and Government policies favouring modern building types and technologies, all contributed to deterioration of the historic fabric. Accordingly, much of the old historic fabric was already demolished before the earthquake.

3.1 Soltani (Imam) Mosque

The Soltani Mosque of Boroujerd was known as Masjed Shah in the Pahlavi Dynasty and today it is called Masjed-e Imam Khomeini. Registered in the National Heritage List (ID number 394), the Imam (formerly Soltani) Mosque has been built in the early Qajar period (circa 1830 AD). Designed after the Imam (formerly Shah) Mosque of Tehran, the monument has been designed as a combination of a Mosque and a theological school with 16 small rooms (hojreh) to accommodate theological students. The faience ornament used in this monument is among the unique samples of Qajar tiles. The older mosque was probably built in 10th century A.D. Soltani means related to Sultan which refers to Fath Ali Shah Qajar who ordered the rebuilding of this building.

4 Damage index

Damage and crack like defects form during the service of structures. The formation of defects is an evolutionary process resulting from the accumulation of damage as a result of the working environment. This damage is usually distributed initially in the joint part or vulnerable locations of structures and only tends to localize into discrete cracks just before failure or when the structure suffers from high loads such as seismic loading. Damage evolution and induced structural deterioration make structural strength weakening obvious so that it becomes vulnerable to failure when these structures suffer from seismic loading. Viewing from the essence of structural failure, each type of failure mode is caused by damage evolution started from defects in materials and developed due to long time of service loading or instantaneous action of disaster loads [6]. Scotta *et al.* [7] proposed two series of indexes, namely Global Damage Indexes, which are representative of the overall structure performance, and Section Damage Indexes, which assess the conditions of reinforced concrete beam-column sections, when they performed the seismic risk assessment by defining compact indexes able to measure the damaging and safety level of structures, also accounting for the economic considerations on buildings rehabilitation. Kratzig [8] proposed damage indicators for estimates of seismic vulnerability by extending structural damage concepts for aging structures to seismic design, where structures fail by passing the damage evolution over an ultimate bound. Shen and Shen [9] established a practical hysteresis model of steel members, which takes the effects of damage accumulation and fracture into consideration, and includes the calculation of damage index, the effects of damage on the yielding point, modulus of elasticity and hardening coefficient of steel, the stress-strain hysteresis relationship and the fracture criterion. In the work by Dimova and Negro [10], the vulnerability of the structure was estimated by a fragility analysis based on fitting the numerical models of the structural response in different seismic intensity levels to the experimental data, in which widely used global damage indices, such as the Park and Ang overall structural damage index [11] were associated with the conditional probability of failure and the damage states of the studied structure were expressed in terms of its fragility.

The development of seismic design criteria based on performance indicators and reliability naturally calls for the description of the probable response and behavior of structural damage under seismic excitations with different intensities that may act on them during their life. Thus, the above methods for estimating the probabilities of failure and the expected damage are developed in terms of global damage indices for a structure. In 2003, Papadopoulos *et al.* [12] introduced an exact, simple and accurate method for the calculation of the damage index which is quicker and simpler than prior methods. In 2005, Colombo and Negro [13] proposed a method for the calculation of the damage index, which has been used irrespective of the material. Lourenco and Roque [2] performed an investigation about the possibility of using simplified methods of analysis and simple indexes as indicators for fast screening and decided to prioritize deeper studies in historical masonry buildings and assess vulnerability to seismic actions. These indexes are based mostly on the in plan dimensions and height of the buildings. The simplified methods indicate that, in Portugal, the average in plain area of earthquake resistant walls and average height are independent of the seismicity.

4.1 Typical damages in monuments

Vulnerability may be reduced through retrofitting, a protection to better resist the seismic demand. Anti-seismic action requires the knowledge of seismic site response, the definition of the seismic load (a rather challenging issue) and the knowledge of the characteristics of existing buildings. A review of the historical sources in Iran, supported by present day knowledge of earthquake engineering, shows that, as far as the earthquake damage is concerned, the slender, free standing members of a complex are in the first line of an earthquake. Minarets, wind towers and high portals of *eivans* and entrance halls are amongst the weakest members. There are numerous references to the collapse of minarets and high portals in past earthquakes. There are in fact very few old minarets, which have not undergone extensive restoration or reconstruction in the areas of higher seismic activities. Remnants of partially toppled minarets, particularly those integrated within the building, are frequently seen in the old mosques. It should be noted that the response of a tall slender structure, such as a minaret, during an earthquake depends primarily on the frequency contents of the ground shaking.

A low frequency, distant shock may easily topple a minaret or a slender tower, but a high frequency local shock may affect the main building more than the minaret. An example of this behaviour is the minaret of *Bagh-i Qushkhane* in Isfahan, the only section of a large building still remaining. High portals of entrance halls and *eivans* have also performed weakly in earthquakes. The upper parts of these portals are effectively, free standing slender elements, susceptible to low frequency ground shaking, in a direction perpendicular to the portal.

Collapse of the upper part of the high portal of *Jami* mosque in Gunabad (built in 13th century AD) during the devastating earthquake of 31 August 1968 (M = 7.4) is a more recent event [14].

4.2 Damage in Imam Mosque

The main damage in the Imam Mosque includes the collapse of the false ceiling including the rich stalactite ornament. This damage is shown in fig. 2. In the retrofitting process increased stiffness creates a negative effect on the building's seismic performance. Other intervention strategies may aim at producing other types of change in the structural behavior, such as increasing the energy dissipation capacity by means of specific devices, or decreasing the inertia forces, for instance by means of base isolation. It is therefore important that the choice of a seismic upgrading strategy considers all the changes in structural behavior it may induce. Moreover, it is very important to know how the solutions adopted influence the seismic resistance of different collapse mechanisms [15]. The destruction shown in fig. 3(a) has resulted from the execution of a big concrete beam over the doorway that changed the behaviour of the same and made it rigid. As is seen, in some cases incorrect retrofitting would lead to structural damage at the time of earthquake. Retrofitting methods should be

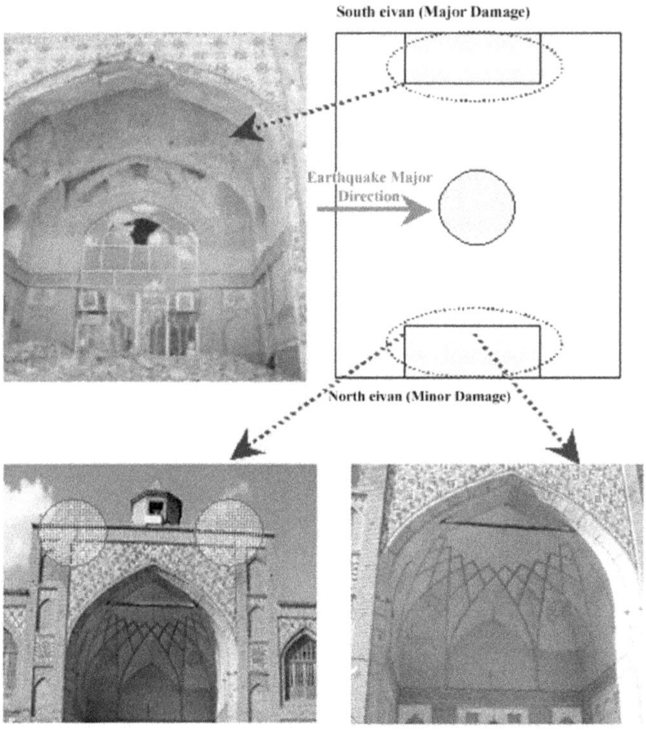

Figure 2: Ornament collapse in the south and north *eivan*.

based on the improvement of the elasticity of the structure and those methods that increase the rigidity of the structure are not suitable solutions for seismic reconstruction. Fig. 3(b) illustrates the seismic performance of this tie beam arch keystone which was destroyed. This matter and the tie beam impact increased damage in the north *eivan*. In the repair process, conductors and consultants must consider seismic loads and they must design and repair with a special brick with a lock system.

(a) The solid concrete tie-beam

(b) Damage in Arch keystone

Figure 3: Ornament collapse in the south and north *eivan*.

The concrete tie-beam on the south *eivan* of Imam Mosque has aggravated the impact of the earthquake on the structure as, during the construction, the keystone of the original arch has also been weakened. The mentioned tie-beam should be replaced by a more resilient system of reinforcement. Fig. 4 indicates the effect of the concrete tie beam on the dynamic behaviour of south *eivan* of Imam Mosque. This beam prevents flexible behaviour and this has caused more damage in the south *eivan* than the north *eivan*.

Figure 4: Effect of concrete tie beam on dynamic behaviour.

Figure 5 shows north *eivan* behaviour against earthquake. This *eivan* had soft, flexible and suitable seismic behaviour. The earthquake has damaged the intermediary structures between the vaults and the finished floor opening the

Figure 5: The north *eivan* after the earthquake.

Figure 6: Water damage after the earthquake.

way for penetration of water. Internal decorations in particular the *mihrab* inscribed in the Imam Mosque are more exposed to water damage. Fig. 6 illustrates water seepage in this monument.

5 Retrofitting process

Besides the seismic intensity at collapse, the method allows the identification of the weakest links and connections in the structure as well as its expected collapse mechanism, which are relevant information to the design of seismic retrofitting solutions. The proposed method is conservative, as it does not account for the energy dissipation capacity, which is likely to be underestimated by means of using an equivalent linear damping coefficient, and overestimates the effects of the seismic action, as it does not account for its duration. It cannot be applied to regular block masonry, as it cannot simulate the behaviour of the interfaces and the geometrical non-linearity. Given the type of mortar used in the monuments, penetration of water can cause several kinds of damage: changing colour; damage to ornament; and increasing structural weaknesses at earthquake time. In Soltani Mosque the damp proof measures to protect the building against the humidity are ineffective and have further aggravated the situation. There is a need to replace the bituminous mats by an appropriate method in line with the traditional water insulation techniques. There is a need to provide temporary light-weight roofing (corrugated sheets) for emergency protection of the roofs against the coming autumn precipitation. This might be used as a general approach for all parts of roof, if time constraints encourage it. Contractors of Imam Mosque involved in repairing the roofing, need to be warned not to dump debris on the roofs. This may cause new visible or invisible damage.

6 Conclusions

Terms of References (ToR) for restoration and consolidation of the monuments should be drafted specifically for each monument according to the particular conditions prevailing in each case. The ToRs of Soltani Mosque should specifically include the following services:

a. Calculating quality and quantity seismic damage index
b. Reducing major risk of structure
c. Investigating dangerous and structural cracks
d. Reducing stiffness of tie beam
e. Damp-proof insulation

Each monument has a protected core zone. In the case of Imam Mosque, the new roofing of Bazaar, with corrugated sheets had destroyed the original fabric of Bazaar and has damaged the urban landscape and assaulted the historic buildings along the bazaar.

In Soltani Mosque severe damage has been incurred by the main rib arches (*tavizeh*) at the front façade of *eivans*. In addition to the earthquake forces the weakness of structures due to the poor use of bricks (jack-arch, instead of vaulting or *rumi*) had aggravated the situation.

Seismic strengthening projects are specialized works, which need a thorough knowledge and experience in this field. It is not a type of work that any novice can easily participate. Therefore, the qualifications of the engineers and contractors should be evaluated and approved by the Committee of Experts as discussed in Item 3 above, and the projects should be assigned to qualified organizations. Of course, the door should remain open for future engineers and contractors to gradually enter this field.

The tie beam effect on the dynamic behaviour of south *eivan* of Soltani Mosque is to prevent flexible behaviour and this has caused more damage in south *eivan* than north *eivan*. Therefore mistakes in the retrofitting process will cause increased damage during an earthquake.

References

[1] Mallardo, V., Malvezzi, R., Milani, E. & Milani, G., Seismic vulnerability of historical masonry buildings: A case study in Ferrara. *Journal of Engineering Structures*, **30**, pp. 2223-2241, 2008.

[2] Lourenco, P. B. & Roque, J. A., Simplified indexes for the seismic vulnerability of ancient masonry buildings. *Journal of Construction and Building Materials*, **20**, pp. 200-208, 2006.

[3] ICOMOS. Recommendations for the analysis, conservation and structural restoration of architectural heritage, 2001.

[4] Vosoughifar, H. R. & Sadat Shokouhi, S. K., Presenting restoration and retrofitting process of Jame' Mosque after Doroud earthquake. *Proceedings of the 4th International Conference on Seismic Retrofitting*, Tabriz, Iran, 2012.

[5] Motamed, J., Adlparvar, M., AlHussaini, A., Vosoughifar, H. R. & Walls, A., Reconstruction and retrofit of the historical monument of Arge Bam after earthquake damage. *The first international congress on seismic retrofitting*, Tehran, Iran, 2006.

[6] Lemaitre, J. & Desmorat, R., *Engineering Damage Mechanics*, Springer-Verlag: Berlin, 2005.

[7] Scotta, R., Tesser, L., Vitaliani, R. & Saetta, A., Global damage indexes for the seismic performance assessment of RC structures. *Earthquake Engineering & Structural Dynamics*, **38(8)**, pp. 1027–1049, 2009.

[8] Kratzig, W. B., Damage indicators for estimates of seismic vulnerability. *Proceedings of the Eighth International Conference on Computational Structures Technology*, Las Palmas, Spain, pp.111–132, 2006.

[9] Shen, Z. Y. & Shen, S., Seismic analysis of tall steel structures with damage cumulation and fracture effects. *Journal of Tongji University*, **30(4)**, pp. 393–398, 2002.

[10] Dimova, S. L. & Negro, P., Seismic assessment of an industrial frame structure designed according to eurocodes, part2: capacity and vulnerability. *Engineering Structures*, **27**, pp. 724–735, 2005.

[11] Park, Y. J. & Ang, A. H. S., Mechanistic seismic damage model for reinforced concrete. *Journal of Structural Engineering, ASCE*, **111(4)**, pp. 722–739, 1985.

[12] Papadopoulos, P., Mitsopoulou, A. & Athanasopoulou, A., Failure indices for RLC building structures. *12th European Conference on Earthquake Engineering*, ed. A. K. Chan, Elsevier, 2002.

[13] Colombo, A. & Negro, P., A damage index of generalized applicability. *Engineering Structures*, **27**, pp. 1164 –1174, 2005.

[14] Maheri, M. R., Seismic vulnerability of post-Islamic monumental buildings in Iran, a review of historical sources. *Fifth International Conference of Seismology and Earthquake Engineering*, Tehran, Iran, 2007.

[15] Bento, R., Lopes, M. & Cardoso, R., Seismic evaluation of old masonry buildings. Part II: analysis of strengthening solutions for a case study. *Engineering Structures*, **27**, pp. 2014–2023, 2005.

Assessment of the earthquake behavior of Hotel Ermionio in Kozani, Greece constructed in 1933 before and after its recent retrofit

G. C. Manos & E. Papanaoum
Department of Civil Engineering,
Aristotle University of Thessaloniki, Greece

Abstract

The earthquake behavior of a retrofitted four-story reinforced concrete (R/C) building, which was constructed in 1933 and belongs to the "Nouveau Monuments of Greece", is investigated here. The predicted earthquake behavior of the old structure, under the provisions of the current Greek Seismic Code, revealed the low resistance of this building to the design seismic loads. The estimate of the strength of existing structural members was based on tests performed in-situ as well as at Aristotle University. The retrofitting scheme included a partial change of the structural system by cast-in-place R/C walls together with the strengthening of all columns and beams by R/C jackets as well as by the use of carbon fiber reinforced laminates (CFRP) for the improvement of the bearing capacity of the slabs. At the same time, all the architectural features of this building were preserved. The predicted earthquake behavior of the retrofitted structure demonstrated its sufficient earthquake resistance.

Keywords: retrofitting, heritage structure, reinforced concrete, CFRP laminate anchorage, flexural behavior.

1 Introduction

This particular building is a 4-story reinforced concrete (R/C) building with a basement located at the central plaza of the city of Kozani in Greece. Its construction commenced in 1931 under the design and supervision of the German engineer Max Ruthven; the owner was a local merchant (K. Vamvakas) and it became the best hotel of the region for a number of years (Hotel

Ermionio). It is representative of the first R/C structures of the region (Western Macedonia) and it has many common architectural features with a similar R/C building that was built at the same period in Nice, France by the same engineer. During the last decade, especially after the strong earthquake sequence of 13 May 1995 that affected the city of Kozani, the building was not in use as it was in need of structural repairs. Furthermore, it could not be demolished and rebuilt since it was protected under the cultural heritage legislation of Greece. The actual design for its structural repair, submitted to the local city planning authorities, is not presented here. Instead, this paper is concerned with a numerical investigation based on the actual geometry and structural details of the building prior to and after its repair. In assessing the expected earthquake performance of this building, use is made of the provisions of the current Greek Seismic Code [1] in finding estimates of the demands imposed on the structural members. The estimates of the corresponding strengths were based on the actual structural details prior to and after its repair as well as on the findings of a specific set of tests that were conducted either in-situ or at the Laboratory. More information is included in the dissertation of the second author [2] as well as in [3].

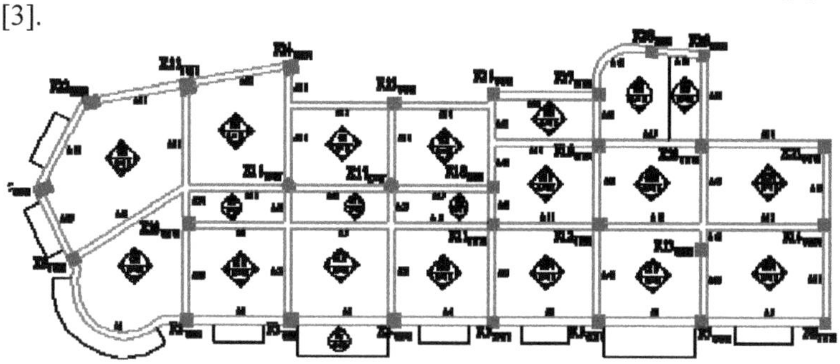

Figure 1: Plan of the typical floor showing the R/C columns, beams and slabs.

2 Assessing the performance prior to the retrofitting

2.1 Level A – quick assessment of the old existing building

This assessment was made according to the guidelines of a relevant assessing method adopted by the Organization of Antiseismic Planning and Protection of Greece [4]. This is a simple method based on the general structural features of a building as well as on particular information about its original design. The grading of this building, according to this Level A assessment was relatively low (total grade = –1 < 0); this is an indication that the building may have a non-satisfactory earthquake performance and one must proceed to level B assessment. The guidelines of the recent Greek Code for retrofitting of R/C buildings [5] include provisions relevant to the various aspects of assessing the expected seismic performance of old R/C buildings prior to retrofitting. These provisions refer to material sampling as well as to the methodology of checking the basic inequality

$$S_d < R_d \tag{1}$$

between the strength R_d and the demand S_d placed upon the various members and the values of the safety factors of either the actions leading to the estimate of the demand or the various parameters leading to the estimate of the strength. This is done for assessing the performance prior to retrofitting as well as for designing and checking a given retrofitting scheme.

This process is broadly followed here although no particular reference is made to specific relevant provisions of this code.

2.2 Level B – assessment of the old existing building

For this assessment level, one needs to get information on the structural details and material properties, based both on the original design files as well as on in-situ information. Moreover, the design assumptions made on the mechanical properties of the materials must be supplemented by information obtained in-situ from the actual structure. To this end, concrete and steel specimens were taken from the slab of the ground floor ceiling and tested at the Laboratory. The average concrete strength found was 11.23 MPa. The corresponding characteristic strength that can be assumed today is C8/10 (f_{ck} = 8 MPa). The reinforcing steel bars were of the smooth surface type with yield stress f_y = 361 MPa for the 8 mm rebar (f_{ult} = 500 MPa) and f_y = 317 MPa for the 10 mm rebar (f_{ult} = 404 MPa) and elongation at fracture larger than 20%. The corresponding yield stress that can be assumed today is f_y = 350 MPa.

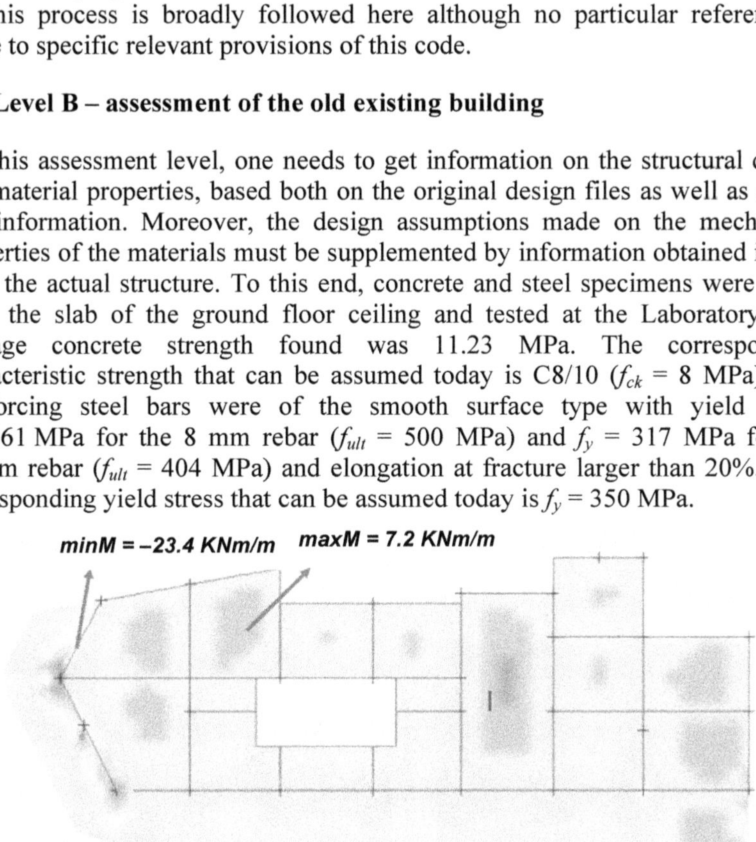

Figure 2: Flexural demands upon the ground floor ceiling.

2.2.1 Slabs

For assessing the expected performance of the slabs, a finite element (FE) analysis was performed for all the slabs with predominant loading the combination of dead and live loads as specified by the Greek Code. Fig. 2 shows the results of the ground floor ceiling that is one of the most-stressed slabs due to the relatively large exterior balconies and internal galleries as well as its live load requirements. The extreme values of some of the demands in terms of bending moments are shown in the same figure. The structural details of this slab are shown in fig. 3.

One piece of the slab (95 mm thick) with dimensions 1.30 m × 0.96 m was removed from the building and was tested under controlled conditions in pure flexure at the Laboratory, resulting in a bending moment capacity $M_{Rd} = 9.18$ kNm/m. The corresponding estimate based on structural details (fig. 3) and relevant calculations is equal to M_{Rd} = 11.78 kNm/m. This non-conservative estimate must be attributed to the long term corrosion that reduced the effective cross-section of the reinforcing bars. Comparing the experimental bending slab strength with the numerical analysis maximum demand (fig. 2) for positive slab flexure, a safety factor equal to 1.27 is found; this indicates an acceptable limit-state performance. However, the demand for negative slab flexure yields a minimum bending moment that is approximately two and a half times larger ($M_{\min}$ = 23.4 kNm/m); this indicates a non-acceptable limit-state performance for the large exterior balconies. The retrofitting addressed this requirement as will be presented in section 3.1.3.

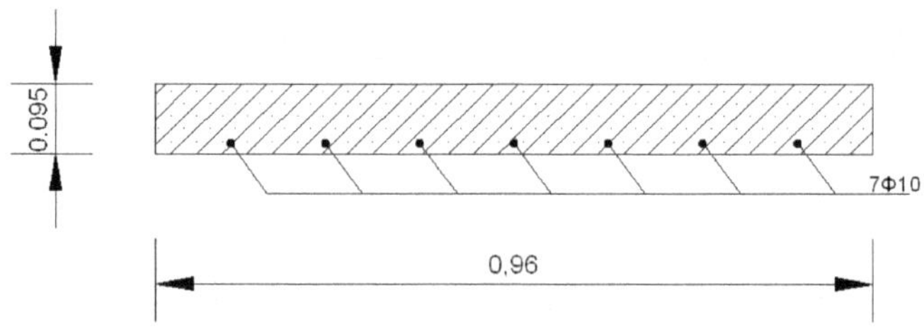

Figure 3: Structural detail of the R/C slab.

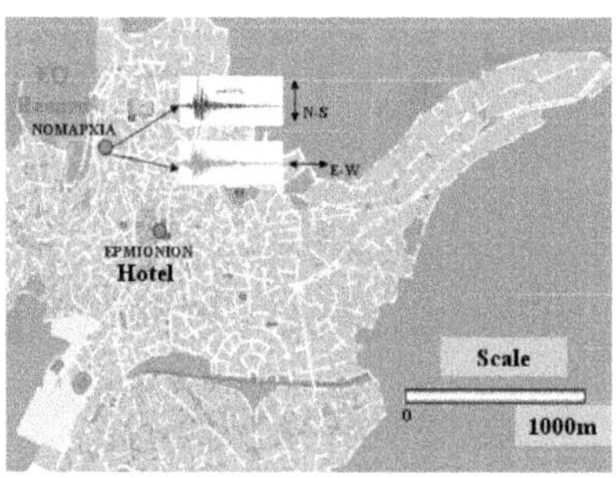

Figure 4: Site of earthquake recording and location of Hotel Ermionio.

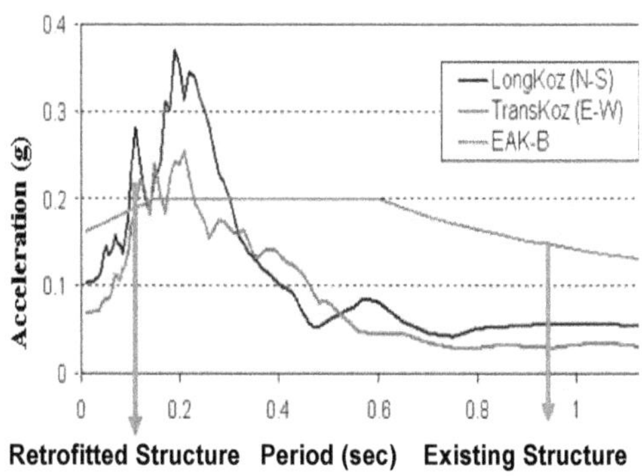

Figure 5: Acceleration response spectrum for the Kozani earthquake of May 1995.

www.witpress.com, ISSN 1743-3533 (on-line)

2.2.2 Multi-story structure

A dynamic numerical analysis was performed for assessing the expected earthquake performance of the multi-story structure according to the provisions of the Greek Seismic Code 2000 [1], using the SAP200 (version 10.1) software. Three different approaches were employed. First, the Simplified Spectral Method, then, the Dynamic Spectral Method based on the design spectrum specified by the Greek Seismic Code and finally, the Dynamic Spectral Method based on the response spectrum that resulted from the 1995 Kozani Earthquake [6-9]. In fig. 5, the design spectrum curve from the Greek Seismic Code (EAK) is included in addition to the acceleration response spectra curves for the two recorded ground acceleration components (East-West, North-South, 5% damping ratio). Moreover, the eigen-periods of the old existing structure as well as those of the retrofitted building are also shown. For the simplified spectral method the following parameters were applied in the numerical analysis:

- Eigen-periods $T_x = 0.47$ s, $T_y = 0.483$ s.
- Seismic Zone I, design ground acceleration 0.16g, soil category B, importance factor = 1, response modification factor $q = 2.0$, damping ratio 5%, response amplification factor $\beta_o = 2.5$, influence of the foundation $\theta = 1$

According to the Greek Seismic Code provisions, the original building is a torsionally sensitive structure with design eccentricities equal to:

$$e_{xmax} = 5.8 \text{ m}, e_{ymax} = 1.25 \text{ m}, e_{xmin} = -3.8 \text{ m}, e_{ymin} = -0.75 \text{ m}.$$

According to the dynamic spectral 3-D method of analysis, the eigen-period values for the original building that are associated with more than 90% of the total mass are listed in the following table.

Table 1: Eigen-period values T_i (s).

T_1	T_2	T_3	T_4	T_5	T_6	T_7	T_8	T_9	T_{10}	T_{11}	T_{12}	T_{13}
1.51	0.96	0.94	0.54	0.33	0.33	0.27	0.16	0.16	0.16	0.12	0.09	0.09

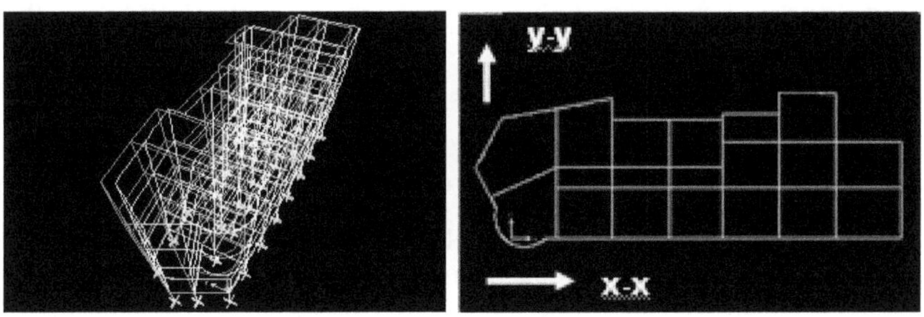

Figure 6: $T_2 = 0.96$ s (70% mass participation along y-y).

From the modal analysis it becomes apparent that there is a degree of coupling between mode 2 and mode 3 with modal participation factors that correspond to nearly 75% of the total mass (see figs. 6 and 7). Fig. 8 depicts the flexural demands for the ground floor columns as they resulted from the various analysis approaches briefly described above.

- The Simplified Spectral Method.
- The Dynamic Spectral Method based on the design spectrum specified by the Greek Seismic Code.
- The Dynamic Spectral Method based on the response spectrum that resulted from the 1995 Kozani Earthquake (figs. 4 and 5).

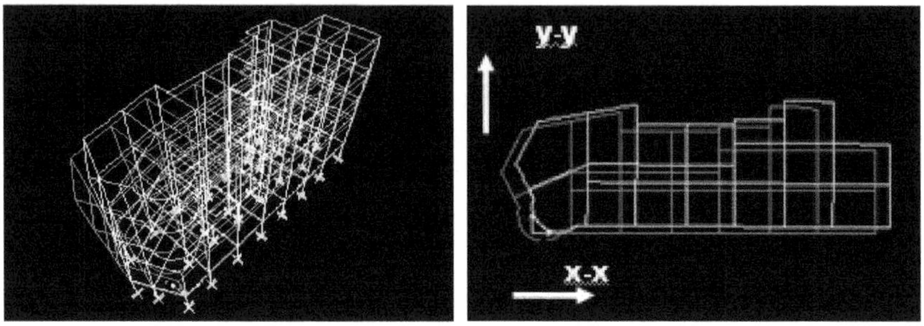

Figure 7: $T_3 = 0.94$ s (70% mass participation along x-x).

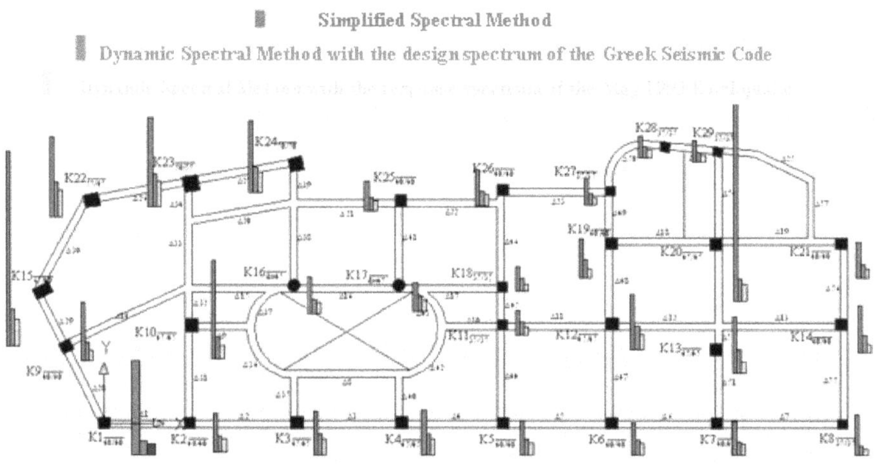

Figure 8: Predictions of the flexural demands for the ground floor columns.

As can be seen from fig. 8, the flexural demands resulting from the Simplified Spectral Method are 2 to 6 times larger than the corresponding demands resulting from the Dynamic Spectral Method based on the design spectrum specified by the Greek Seismic Code. This must be attributed to the torsional sensitivity of the existing old structure, which is penalized by the provisions of the Greek Seismic Code when the Simplified Spectral Method is applied. When the results of the Dynamic Spectral Method based on the response spectrum of the 1995 earthquake are compared with the corresponding results based on the design spectrum included in the Greek Seismic Code it can be concluded that the former are 1.25 to 1.75 times smaller than the latter. This must be attributed to the difference in the amplitude of the corresponding design spectral curves when combined with the dominant eigen-period values of the existing old building (see fig. 5). Table 2 lists the predicted inter-story drift values, in terms of angular inter-story response, for the various stories of the old existing structure as predicted by the Dynamic Spectral Method based on the design spectrum specified by the Greek Seismic Code. As can be seen these inter-story drift values are more than 4 times larger than the limit of 0.005 set by the Greek Seismic Code.

Table 2: Inter-story drift values for the frame K1-K9-K15-K22 in the South.

Angular Inter-story Drift	Ground floor (h=4.09 m)	Mezzanine (h=2.60 m)	1st story (h=3.60 m)	2nd story (h=3.75 m)
γ	**0.0233**	**0.0226**	**0.0219**	**0.0166**

When the flexural strength capacities of the columns of the old existing building are compared with the flexural demands, predicted as explained above, the demands by far exceed the capacities when these demands are based on the Simplified Spectral Method; when these demands are obtained from the Dynamic Spectral Method based on the design spectrum specified by the Greek Seismic Code they exceed the flexural capacities of almost all the Ground Floor columns. The flexural capacities satisfy the demands for the columns of the structure only when the predictions are by the Dynamic Spectral Method and the response spectrum of the 1995 Kozani Earthquake acceleration recording is used. As pointed out, this is due to the frequency content of this particular ground motion combined with the eigen-frequencies of the old existing structure (figs. 4 and 5)

3 Structural retrofitting

The structural retrofitting (fig. 9), aimed to upgrade the earthquake performance of the building, included jacketing of all the columns and beams (with a thickness of 75 mm for the columns and 50 mm for the beams) as well as the construction of shear-walls (with a thickness of 300 mm), within the bays of the old structure at the North and East perimeter of the building that would not obstruct its functioning which could be served from the other two sides (South and West) facing the main square of the city. The structural retrofitting scheme applied in this particular case combines the following two basic structural retrofitting principles:

a) To increase the strength and ductility of individual structural members by the construction of R/C jackets around the columns and beams in a way that the retrofitted members' strength can successfully meet the seismic demands imposed on the retrofitted structural members.
b) To partially change the structural system by transforming some of the frames at the back sides of the building to R/C shear walls or by alternative changes

in the structural system. In this way, one aims at decreasing the seismic demands on the individual structural members.

In this way, one tries to vary the terms of both sides of inequality (1) so that a favourable final performance results for all the structural members of the retrofitted structure. The provisions of the Greek Code for design of the repair and strengthening of R/C structures [5] allows the cast-in-place R/C wall, when its thickness is less than the thickness of the surrounding members, to be encased within the bay of the surrounding frame either non-connected or partially connected to the top and bottom beam of the surrounding frame. It also allows, when the thickness of the cast-in-place R/C wall is larger than the thickness of the members of the surrounding frame, to construct a fully reinforced shear wall that encases all the frame's members. For this type of cast-in-place R/C walls the following three distinct cases are provided for:

1. The cast-in-place R/C wall does not have any particular connection with the members of the surrounding frame apart from simple contact.
2. The cast-in-place R/C wall is connected only at the top and the bottom with the corresponding beams of the surrounding frame. This connection is done with steel anchors placed at certain intervals.
3. The cast-in-place R/C wall is connected at the top and the bottom with the corresponding beams as well as left and right with the corresponding columns of the surrounding frame. This connection is also done with steel anchors placed at certain intervals.

It should be mentioned here that one could utilize either one of the above a) and b) structural retrofitting principles on its own. Thus, one could try to form a structural retrofitting scheme that would rely only on upgrading the strength of the individual structural members of this building by jacketing or other means. Alternatively, one could introduce changes to the structural system, by adding shear walls or other means, in an effort to bring the seismic demands to acceptable levels with the existing strengths of the structural members. The structural retrofitting scheme, that is presented in this paper, combines both the structural retrofitting principles, described above as a) and b), in order to achieve the desired goal; despite this conservative design combination, this structural retrofitting in cost terms represents a relatively small part of all structural, non-structural and maintenance cost for works made on this building.

As this is a listed building, belonging to the Nouveau Monuments of Greece, all the non-structural elements of the façade (door and window openings, wooden shades, cast iron railings etc.) as well as selected non-structural elements in the interior had to be preserved.

As already mentioned, this structural retrofitting scheme resulted in both increasing the torsional stiffness of the structure and thus decreasing the seismic demands as well as upgrading the capacities of all the columns and beams; it extended all the way from the foundation to the top of the building. The specified retrofitting materials are C20/25 for the concrete and S500s for the steel. Moreover, the structural retrofitting included the upgrading of the flexural capacity of the slabs by applying CRFP laminates (see section 3.13).

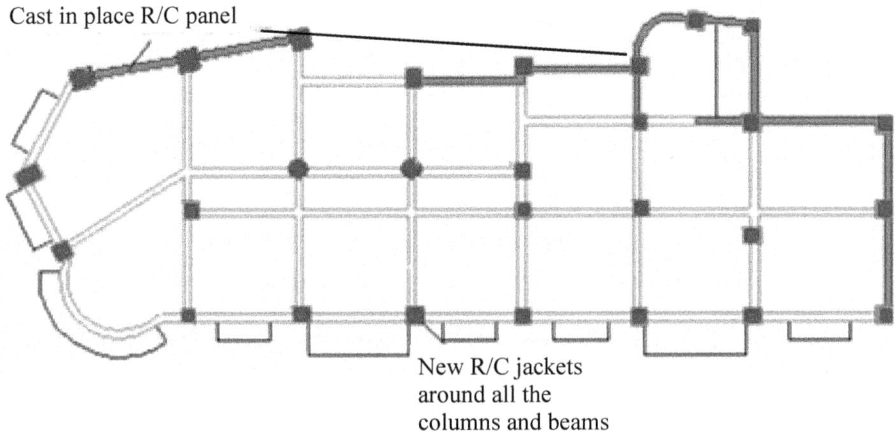

Figure 9: Studied retrofitting scheme.

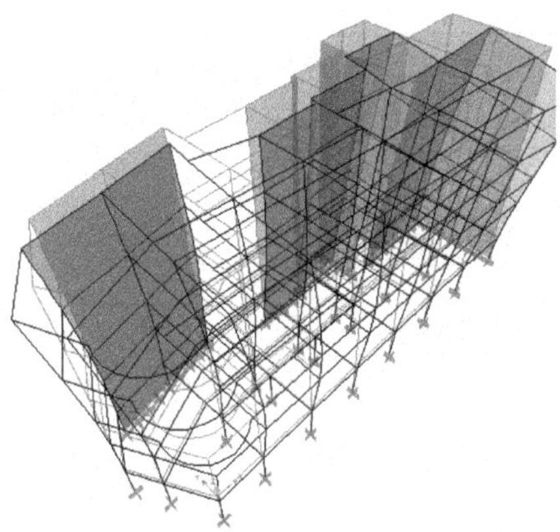

Figure 10a : $T_2 = 0.15$ s (54% mass participation along *y-y*).

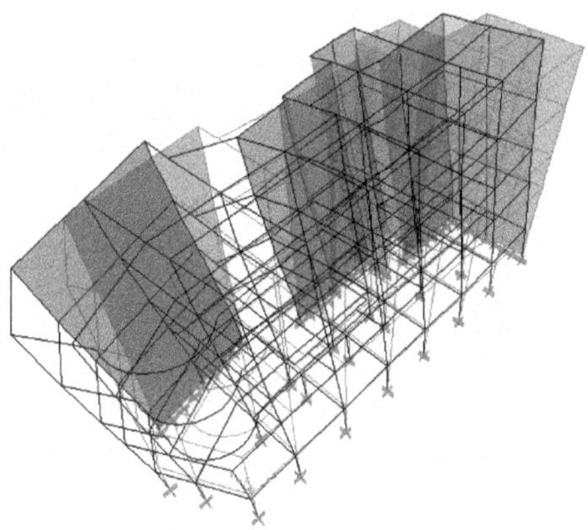

Figure 10b : $T_4 = 0.09$ s (56% mass participation along *x-x*).

3.1 Level B assessment of the retrofitted building

The expected earthquake performance of the retrofitted structure was assessed in the same way as described in section 2.2.2 employing the previously described methods of analysis and numerical tools. In what follows, a summary of this assessment will be presented by comparing critical response parameters of the retrofitted building with those of the existing old building.

3.1.1 Influence of the retrofitting on the eigen-modes and eigen-periods

In figs. 10a and 10b the dominant eigen-modes of the retrofitted structure are shown together with the corresponding eigen-periods, which are also listed in table 3. As can be seen, the retrofitting scheme resulted in a significant increase of the lateral stiffness of the structure.

Table 3: Comparison of eigen-period values between old and retrofitted structure

Structural scheme	Eigen-period for dominant *y-y* response	Eigen-period for dominant *x-x* response
Old Structure	$T_2 = 0.96$ s	$T_3 = 0.94$ s
Retrofited Structure	$T_2 = 0.15$ s	$T_4 = 0.09$ s

3.1.2 Influence of the retrofitting scheme on the maximum displacement and stress response

The influence of the retrofitting scheme is studied by comparing the maximum response values, in terms of displacements and stress-resultant values relevant to flexural response of the columns, by applying the Dynamic Spectral Method based on the design spectrum specified by the Greek Seismic Code [1].

Table 4: Comparison of displacement response between old and retrofitted structure.

Top Story Corner	Old Existing Building			Retrofitted Structure		
	u_x (mm)	u_y (mm)	θ_z (rad)	u_x (mm)	u_y (mm)	θ_z (rad)
S-W	51.36	56.99	$3.64 \cdot 10^{-4}$	0.62	5.79	$1.33 \cdot 10^{-4}$
N-W	51.10	47.37	$5.87 \cdot 10^{-4}$	0.95	1.50	$2.47 \cdot 10^{-4}$
N-E	50.34	47.37	$7.08 \cdot 10^{-4}$	2.51	1.50	$3.57 \cdot 10^{-4}$
S-E	50.30	56.51	$8.28 \cdot 10^{-4}$	2.64	5.54	$3.45 \cdot 10^{-4}$

As can be seen in table 4, the increase in the lateral stiffness of the structure resulted in a drastic decrease of the horizontal displacement response at the ceiling of the third story (top of the building). This decrease occurred despite a relative increase in the lateral seismic forces resulting from the increase of the spectral values corresponding to the dominant eigen-modes of the retrofitted structure when compared with those of the old existing structure (fig. 4). This is also true for the stress-resultant response values relevant to flexure of the columns. These maximum response values are listed in table 5 for all the columns of the Ground Floor before and after retrofitting. As can be seen this maximum response is reduced by 2 to 4 times as a consequence of the applied retrofitting scheme.

A critical element in the applied retrofitting is the strengthening of the existing foundation. The positive aspects regarding the retrofitting of the existing foundation were first, that it was above the water level and second, that it was accessible all around the external perimeter of the building; thus it could be easily retrofitted by new external R/C jackets. Thus, new external R/C jackets were constructed at the top of all the existing footings together with new cast-in-place R/C beams springing from each footing and connecting all the adjacent columns. These new cast-in-place R/C foundation beams were also connected to the reinforcing cages that formed the jacketing of all the corresponding columns

Table 5: Comparison of stress–response between old and retrofitted structure.

Ground Floor Column No	Old Existing Building			Retrofitted Structure		
	N (kN)	M_2 (kN·m)	M_3 (kN·m)	N (kN)	M_2 (kN·m)	M_3 (kN·m)
266	89.92	108.91	121.09	30.48	51.39	38.11
267	246.90	104.47	131.70	73.32	36.57	24.51
279	180.76	103.21	134.29	38.03	16.79	25.93
287	269.94	114.20	120.19	53.06	5.58	25.29
291	197.96	76.08	64.14	102.71	4.17	10.69
299	311.10	150.70	79.06	151.79	71.05	3.65
323	74.65	141.96	87.03	25.82	1.24	3.46
327	317.95	166.11	297.42	64.46	71.53	35.21
347	81.33	182.72	173.28	2.97	0.06	2.31
351	205.96	107.81	92.76	211.63	13.02	9.30
266	89.92	108.91	121.09	30.48	51.39	38.11

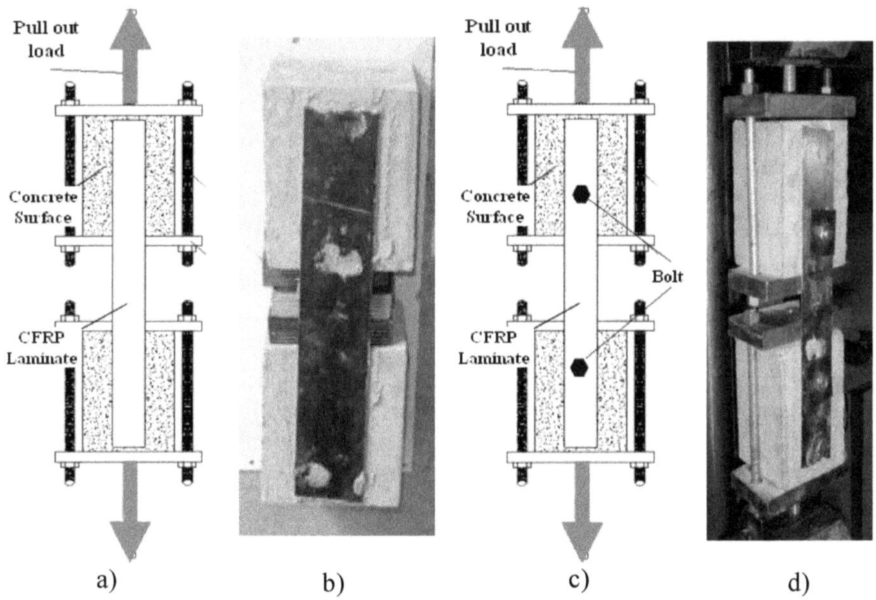

Figure 11: Bond tests of the used CFRP laminates.

starting from the foundation level and extending all the way to the top of the building. The reduced flexural demands for the columns, resulting from the cast-in-place R/C walls at the back sides of the building all along its height, combined with the increased capacities of the structural elements due to the construction of new R/C jackets for all the structural elements (columns and beams) lead to a satisfactory earthquake performance of this building after retrofitting.

3.1.3 Upgrading the flexural capacity of the slabs

As described in section 2.2.1, the assessment of the expected performance of the slabs revealed that they were in need of being upgraded regarding their flexural capacity. Towards this objective, a retrofitting scheme needed to be applied to these slabs. This scheme was planned to make use of a certain type of CFRP laminate with a cross-section of 50 mm × 1.2 mm. These laminates could be applied in-situ either on the upper or lower surface of the concrete slabs with the use of a special epoxy paste. An extensive experimental parametric study was performed on this type of attachment by testing a series of specimens prior to applying the best attachment detail to prototype slab specimens. The tests were performed at the laboratory of Strength of Materials of Aristotle University and utilized twin prismatic concrete specimens with dimensions 100 mm × 100 mm × 150 mm each. These twin concrete prisms were joined with the CFRP laminates; they were attached at the two opposite sides of each twin concrete prism. These specimens were then subjected to such a loading that would force the laminates to be detached from the concrete surface. This is shown in figs. 11a to 11d. A variety of attachments between the CFRP laminates and the twin concrete prisms were tried. The simplest form of attachment was the one employing only epoxy paste between the CFRP laminates and the concrete surface (figs. 11a and 11b). Next, a variety of bolting arrangements were utilized with or without the epoxy paste. Figs. 11c and 11d depict such an attachment of the laminates whereby the epoxy paste is combined with one bolt on each side of the twin concrete prism. The bond strength in the first case was found equal to 2.70 MPa whereas in the second case, it was equal to 4.40 MPa, which represents an increase of 63%. The maximum bond strength that was achieved throughout these tests reached the value of 7.20 MPa, which represents an increase of 167% from the simple attachment of the laminates only with epoxy paste. These findings were utilized with slab specimens that were taken from the actual mezzanine slab of the structure; they were cut from parts of the slab where an opening would be formed for a new staircase.

Figure 12 depicts the loading arrangement that was used to subject these slab specimens to four-point flexure. Initially, this was done for a specimen without any retrofitting that reached a maximum bending moment value equal to 8.84 kNm. Then two laminate strips, each having a cross section of 50 mm × 1.2 mm, were attached on the bottom surface of this slab specimen, which was reloaded to flexure as before. Fig. 13 depicts the obtained flexural behavior of this retrofitted slab specimen. At this stage the CFRP laminates were attached to the specimen only with the use of epoxy paste. This time the specimen reached a maximum bending moment value equal to 20.89 kNm, which represents an increase equal to 136% when compared to the maximum bending moment value of the un-retrofitted slab. This retrofitted slab failed in the form of debonding CFRP laminates (fig. 14). The debonding of the two CFRP laminates was followed by a sharp decrease in the flexural bearing capacity together with the formation of a plastic hinge at mid-span in the form of excessive plastic deformations of the old steel reinforcing bars at the bottom of the slab as well as crashing of the concrete at the top of the slab.

Figure 12: Loading arrangement of the slab specimens in flexure.

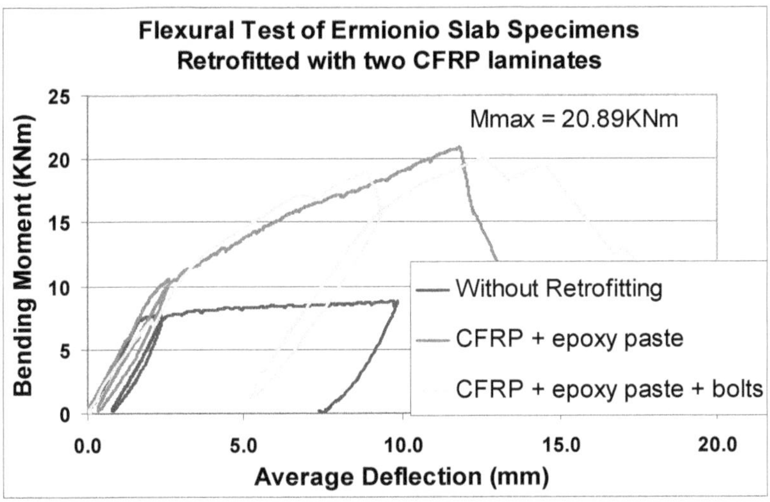

Figure 13: Flexural behavior of Ermionio slab specimen retrofitted with two CFRP laminates either bonded only with epoxy paste or with epoxy paste and bolts.

Next, two new CFRP laminates were reattached to the same slab specimen; this time epoxy paste was used again together with bolts penetrating through the slab and securing the attachment of these CFRP laminates. The flexural bearing capacity of the slab this time reached the same maximum bending moment value as the one observed when only epoxy paste was used for the attachment of the laminates; they exhibited satisfactory performance without any signs of failure either in the form of debonding or fracture. This is shown by the bending moment versus mid-span deflection curve plotted for this test in fig. 13. Despite the satisfactory performance of the CFRP laminates, no increase in the flexural bending capacity could be achieved due to the slab damage from its previous loading history; however; the performance of the anchors for the attachment of the CFRP laminates was satisfactory and the flexural behavior of the slab exhibited a much larger range of deformability (fig. 13).

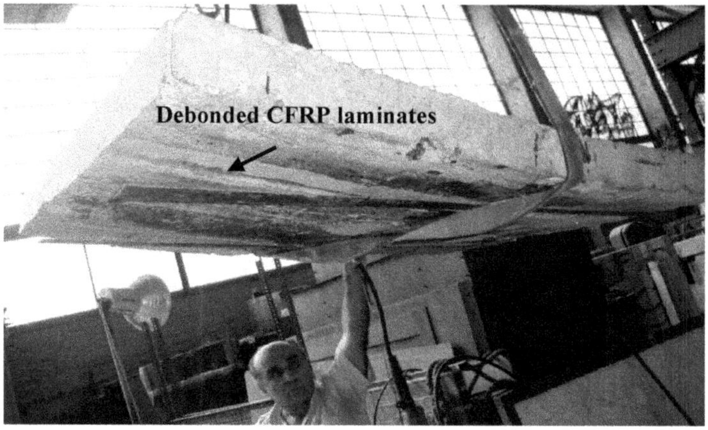

Figure 14: Flexural failure of the Ermionio slab specimen retrofitted with two CFRP laminates bonded only with epoxy paste. The CFRP laminates were debonded from the concrete slab in this case.

4 Conclusions

The detailed assessment of the old existing building revealed that an unsafe earthquake performance could be expected, especially for the Ground Floor columns as well as for parts of the slab of the Ground Floor ceiling.

It is demonstrated that the retrofitting scheme actually applied to this structure, which combines a partial change of the structural system by cast-in-place R/C walls together with new R/C jackets for all the columns and beams, is successful in meeting the demands specified by the current Seismic Code of Greece. At the same time, all the architectural features of this building, which belongs to the Nouveau Monuments of Greece, were preserved.

This retrofitting scheme was supplemented with a scheme to upgrade the flexural capacity of the slabs by applying a certain type of CFRP laminates; the increase in the flexural capability of the slabs of this building resulting from such retrofitting scheme was validated by a special laboratory investigation.

The successfully retrofitted building is already in use again as a luxury hotel.

References

[1] Provisions of Greek Seismic Code 2000, OASP, Athens, December 1999.
[2] Papanaoum, E., Estimate of the earthquake performance of a 4-story R/C building before and after retrofitting, Department of Civil Engineering, Aristotle University, 2006.
[3] Manos, G.C. & Papanaoum, E. Earthquake behavior of an R/C building constructed in 1933 before and after its repair. *STREMAH XI*, ed. C.A. Brebbia, WIT Press: Southampton, pp. 465-475, 2011.
[4] Organization of Earthquake Planning and Protection of Greece (OASP), Guidelines for Level – An earthquake performance checking of buildings of public occupancy, Athens 2001.
[5] Organization of Earthquake Planning and Protection of Greece (OASP), Guidelines for Retrofitting in Reinforced Concrete Buildings, Athens 2011.
[6] Manos G. C., Consequences on the urban environment in Greece related to the recent intense earthquake activity. *International Journal of Civil Engineering and Architecture*, **5**(**12**), pp. 1065–1090, 2011.
[7] Hatzfeld, D., Karakostas, P., Ziazia, M., Selvaggi, G., Berge, C. & Makropoulos, K., The Kozani-Grevena (Greece) earthquake of May, 13th 1995, a seismological study. *Journal of Geodynamics*, **26**(**2-4**), pp. 245-254, 1998.
[8] Theodoulidis, N. & Lekidis, V., The Kozani-Grevena, Northern Greece, earthquake of May 13, 1995: strong motion data and structural response. *Journal of European Earthquake Engineering*, **1**, pp. 3-13, 1996.
[9] European Strong Motion Database, the Kozani Earthquake, 1995.

Seismic strengthening of churches as a part of earthquake renewal in the Posočje region, Slovenia

M. Uranjek, B. Dolinšek & S. Gostič
Building and Civil Engineering Institute ZRMK, Slovenia

Abstract

Seismic activity of the Posočje region, located in the north-east of Slovenia is considerably high. Four major earthquakes have hit this area in the last three decades. Among buildings to be repaired and/or strengthened, churches represent a significant share of the building fund. Churches, which are often classified as cultural heritage, are usually an indispensable part of settlements and their identity. In order to preserve these prominent buildings for future generations, there is a need for attentive and consistent planning of preliminary research, strengthening procedures and their execution. If compared to ordinary buildings in the Posočje region, which are typically characterized by inadequate building quality, churches in this region are of better quality. Nevertheless, on account of their typical design, strengthening procedures are more complex and demanding, and therefore, adequate seismic resistance is harder to achieve compared to ordinary buildings. In the paper, some of the basic strengthening procedures applied on churches in the Posočje region are presented.
Keywords: church, strengthening, earthquake renewal, seismic resistance.

1 Introduction

Seismic activity of the Posočje region, located in the north-east of Slovenia is considerably high; namely four major earthquakes hit this area in last three decades. The first earthquake in May 1976 reached the grade IX-X and the second one in September 1976 the grade of IX according to the EMS scale. Earthquakes in 1976 caused damage to around 12,000 buildings, of which 4,200 were completely destroyed. Consequently people in this area became aware of the danger and begun to construct buildings, which were safer and more resistant to seismic actions. In 1998, an earthquake reaching the grade of VII-VIII damaged 4,000 buildings and destroyed 500 of them. The latest earthquake in 2004 with the grade of VI-VII caused damage to 1,800 buildings and destroyed 20 of them. Among buildings that have so far been repaired or some of the strengthening procedures were carried out, there are well over 150 buildings of cultural heritage, 54 of which churches. Because of their typical structural system, which usually has to remain intact, the selection of appropriate strengthening measures is a complex and demanding process. Therefore it is in most cases harder to achieve adequate seismic resistance for churches than for ordinary buildings. Radical strengthening measures such as new reinforced concrete frames or even additional load-bearing walls, which can be used to improve earthquake resistance of residential buildings, are not eligible for churches. In case of churches, adequate load bearing capacity of separate structural elements and the integrity of the structure as a whole has to be achieved without visible (or distracting) modifications of the basic structural system.

2 Typical structural design and earthquake damage of churches

Although churches were mostly built by experience without detailed calculations, structural elements such as "sewn" corners of walls and iron wall ties (fig. 1) are proof that the builders of the past were familiar with the principles of improving the structural integrity of buildings subjected to seismic loads.

Figure 1: Sewn corners of a St Michael church in Ljubinj and iron wall ties of the church of St Andrew in Svino.

Churches are characterized by a complex structural system constructed out of several structural parts with different geometry and mechanical properties. Consequently, structural parts such as portal, nave, presbytery, bell tower (usually connected to the nave, seldom self supporting) and sacristy, have rather different response to seismic action. Nave as a central structural part of a church is characterised with high walls, which are often not properly connected and anchored at floor and roof levels and in addition, most naves have no load-bearing walls in transverse direction, which makes them irregular in terms of earthquake resistant design. The irregularity of bell towers as a second most discernible part of a structural system is mainly due to its height/floor area ratio. Damage induced by an earthquake often occurs at connections and intersections of different structural parts of the church building. Usually the first elements to suffer the earthquake damage are heavy console elements e.g. campanulas and decorative elements such as jutting roof, especially if not properly anchored and connected to the base structure. Due to movements and oscillations of supporting walls, cracks are often formed in arches and vaults, but also in supporting walls themselves. In fig. 2, damage that occurred on bell tower of a church of St Anton the Hermit and the console structure of a historical building in Bovec is presented.

Figure 2: Typical damage on bell tower of a church and console structure of a historical building.

3 Preliminary investigation

In order to maximize the accuracy of assessing the morphology, damage and mechanical properties of a structure, several methods of investigation should be used. A combination of non-destructive (NDT) and minor destructive (MDT) investigation techniques has proven to be very efficient. NDT techniques such as ground penetrating radar (GPR) measurement and pulse sonic tests can be performed on a wider area of the wall, while MDT techniques (surface and depth sounding, coring, flat jack test), are preferred to be done locally in order to minimize the damage. In some rare cases (because of the complexity and considerable costs) destructive (DT) investigation techniques may be performed.

3.1 NDT investigation techniques

GPR measurements can be used to assess the morphology of the structure. The method is based on the emission and reflection of very short electromagnetic impulses by an antenna system. Reflection of the emitted impulses occurs at the interfaces between materials with different permittivities or conductivities, such as the interface between air (i.e. the voids) and the building material. Because the propagation velocity and the signal penetration depend on the electric and dielectric properties, the boundaries between different materials can be distinguished. GPR can be performed before and after grout injection to evaluate the quality of this strengthening technique (fig. 3).

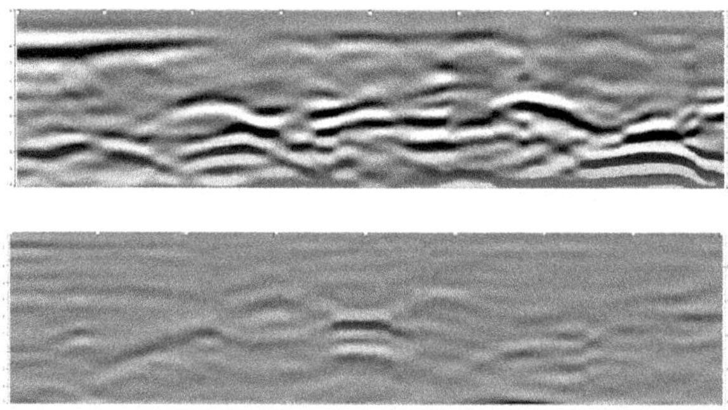

Figure 3: GPR profiles before (above) and after (below) grout injection.

Thermo-graphic measurements are usually used for detecting hidden mistakes in building envelopes, but the method can be successfully applied in the monitoring of the grout injection quality (fig. 4). Thermo-graphical measurements enable us to identify the wall areas with increased surface

Figure 4: Digital and thermo-graphic photography of the wall 28 hours after grouting.

temperature during a given time interval after injection when, because of the hydration process, heat is released. It has to be emphasized that the measured temperature of the wall depends not only on the hydration but also on the external temperature. This is why the results need to be interpreted with knowledge of the external temperatures at the time of measurement, and a comparison between the measured temperatures of the injected and non-injected parts of the wall.

A sonic pulse test is based on the generation of sonic impulses at a selected point of the structure. Based on the time that the impulse takes to cover the distance between the transmitter and the receiver, the quality and homogeneity of the tested structure can be determined.

3.2 MDT investigation techniques

MDT investigation techniques such as surface and in-depth probing (fig. 5) and coring (fig. 6) on selected areas of the investigated building's structure (walls, foundations), reveal the morphology of the structure at tested positions.

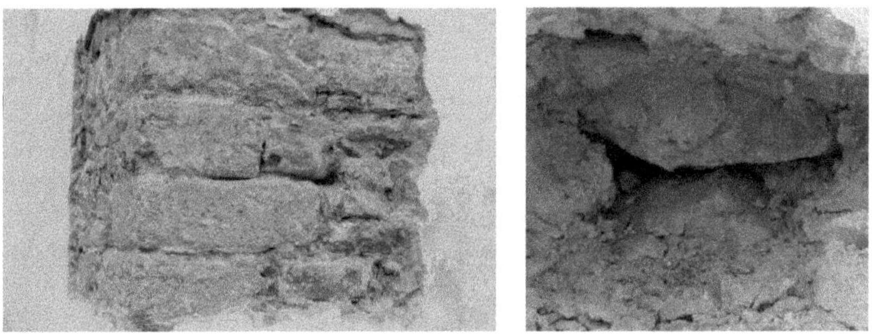

Figure 5: Surface and in-depth probing of the walls.

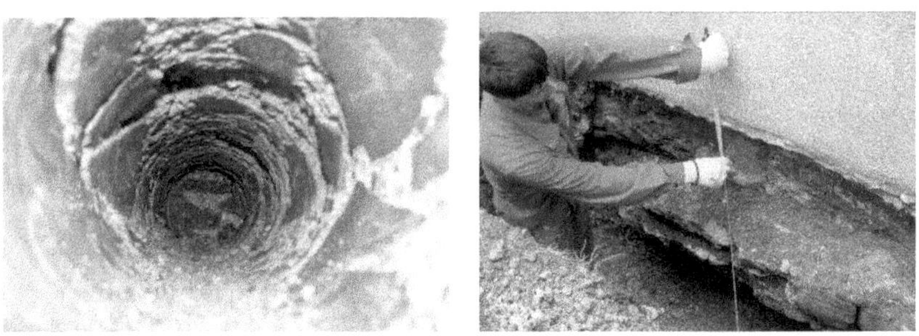

Figure 6: Coring of the walls and in-depth probing of foundations.

Occasionally, more demanding MDT methods, such as the double flat jack test, are used. Flat jack tests enable the determination of the deformability characteristics of the tested wall area. At first, two parallel cuts are made, after which a thin flat-jack is placed inside each of the cuts and the oil pressure operating the jack is gradually increased. LVDT's placed between the two cuts enable the monitoring of vertical and lateral deformations during the test (fig. 7).

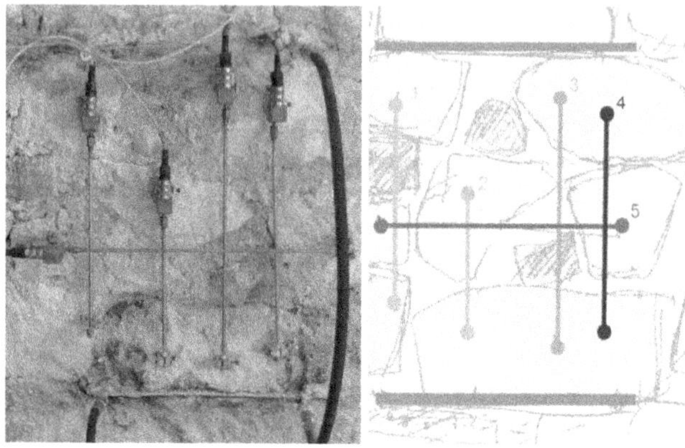

Figure 7: Set up of the measuring equipment for the double flat jack test.

3.3 DT investigation techniques

DT tests are usually carried out in-situ, although it is possible to cut out parts of a structure, carefully transport them and proceed with the tests in the laboratory. Among DT investigation techniques in Slovenia, several in-situ shear tests were performed, usually on fixed-end wall specimens. Such specimens are loaded laterally by means of a hydraulic actuator by increasing the displacement up to the limit state. In-situ shear tests enable the determination of the shear resistance to lateral load and effective stiffness of the tested walls.

4 Evaluation of seismic resistance

Generally the seismic resistance of existing buildings is verified twice: in their existing state and state after strengthening. Seismic analysis of the building in its existing state reveals critical areas, which have to be strengthened, while the basic aim of "post-strengthening" analysis is to determine the level of seismic resistance reached by the strengthened structure. For most churches under earthquake renewal in Posočje, the seismic resistance was evaluated by means of non-linear static analysis using the pushover method, based on the storey-mechanism approach. Based on the idealized bi-linear hysteresis envelopes of the individual walls, the storey resistance envelope is calculated step-by-step by increasing the imposed displacements. In addition to other assumptions, the method assumes that the floors are rigid enough to allow distribution of seismic forces onto the walls and that the first vibration mode shape is the predominant one. The method can quite accurately describe the behaviour of simple masonry buildings with regular distribution of load bearing walls in both horizontal directions and adequate configuration of building in elevation. However, in case of more complex structural systems such as churches, with irregular structural layout and flexible wooden floors, the drawbacks of the method become more obvious. Therefore, in order to determine the critical failure mechanism of the structure during an earthquake, the method can be used by performing seismic analysis at different levels, depending on geometric characteristics of the structure (fig. 8). The division of structure into macro elements correspond to the observed crack patterns after an earthquake.

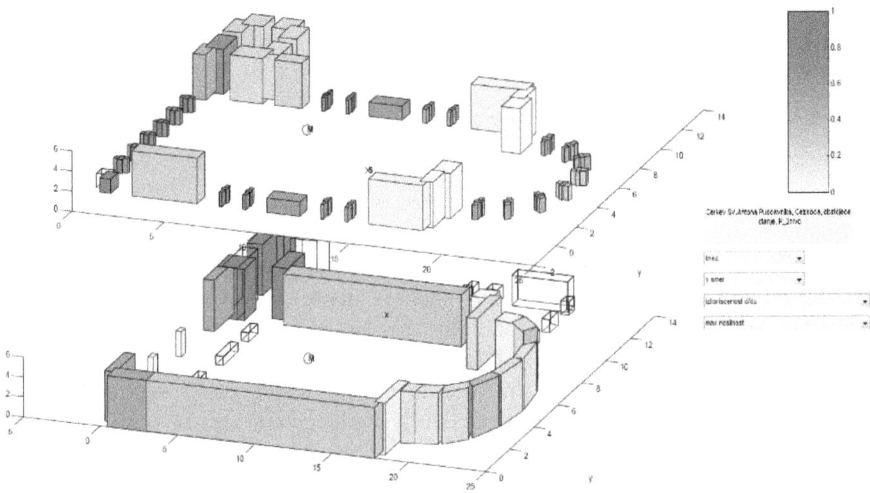

Figure 8: Actual/ultimate displacement ratio for limit state of a structure at two levels.

Recent research [1] has shown that for complex structures, such as churches, more realistic results can be obtained by using the frame by macro elements (FME) method, which takes into account the whole building response instead of just single storey response, and furthermore, allows modal analysis to be performed.

5 Strengthening procedures

Next, some of practically implemented strengthening procedures under post-earthquake renewal in the Posočje region are presented. There are also other strengthening procedures such as sewing of walls with grouted anchors, application of polymer grids or carbon fibre-reinforced polymer wraps that will not be described in this paper.

Tying walls with steel ties (fig. 9) can be applied at the level of individual inter-storey floor structures. In this way the integrity of a structure is improved, the horizontal load is distributed to the walls according to their stiffness and the walls are better protected against excessive rocking and possible failure in the out-of-plane direction.

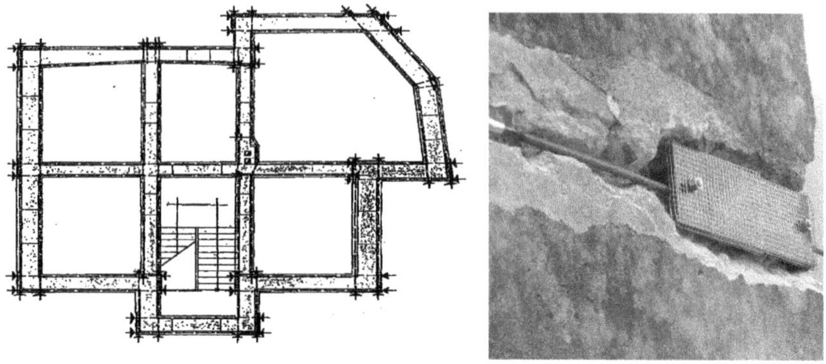

Figure 9: Tying of walls with steel ties.

Reinforced concrete tie-beams can be constructed at floor or roof level and anchored onto the load-bearing walls. Such a method of strengthening improves structural integrity of a building and ensures more uniform behaviour of the structure. By anchoring the roof into the tie-beams possible uncontrolled movements of the roof elements and failure of head-walls is prevented.

Injected anchors applied at the middle of the wall's cross section can, if properly designed and executed, successfully replace steel ties or reinforced tie-beams. This solution is similar to the method used by the old builders. More specifically, connecting walls with ties, installed in the middle of the wall, has been prescribed even after the 1895 earthquake in Ljubljana [2]. Injected anchors can perform either as untensioned steel reinforcement or as prestressed tendons. The execution of a borehole and examples of application are shown on fig. 10.

Figure 10: Application of injected anchors [3].

Repointing can be applied where bed-joints are relatively level, the mortar is poor and units are good. The resistance of a wall to vertical and lateral loading can be considerably improved by replacing a part of the existing mortar with mortar of better quality [4]. Although repointing is usually applied on brick

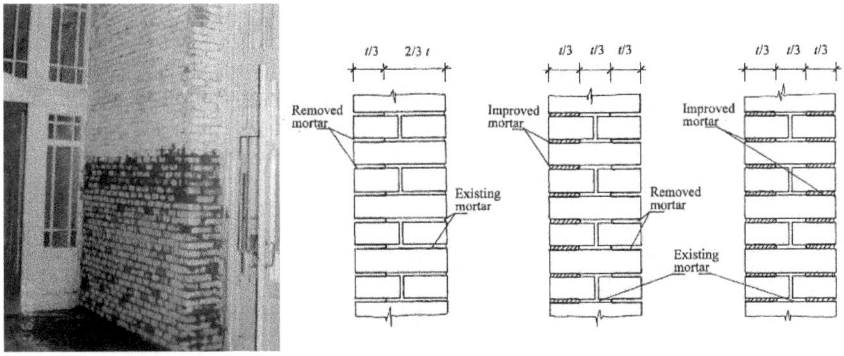

Figure 11: Repointing of a brick-masonry wall [4].

masonry, some authors [5] propose the application of the procedure also on stone masonry walls.

Injection grouting can be used to improve the mechanical properties of two or three-leaf stone masonry walls. Because of inadequate connections between separate stones, voids and poor quality of used mortar, the load bearing capacity of such walls, especially towards horizontal loading, is not sufficient. With grout injection technique, grout (liquid mixture made of water, binder and additives) is injected into a masonry wall (fig. 12) under moderate pressure (2-3 bars). The aim of the technique is to achieve better bonding between separate stones and leafs after the hardening of the injected material and consequently to improve mechanical properties of the masonry. Work on site showed, that it is best to carry out injection grouting of all load-bearing walls, over their whole height of the structure. Damage that occurred in the buildings after strengthening and repeated earthquake has shown that grouting only parts of walls is not a good solution.

Strengthening of walls by applying reinforced cement coating (fig. 13) is usually used for strengthening of brick masonry. However, for more vulnerable structures such as bell towers, it can also be used for strengthening the stone

Figure 12: Injection grouting.

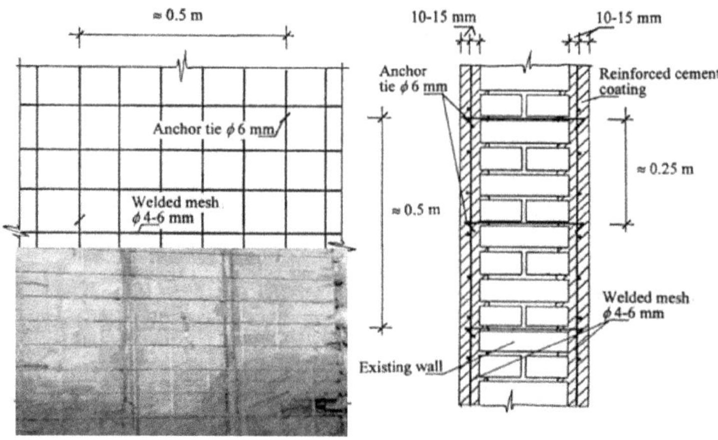

Figure 13: Application of reinforced cement coating on brick-masonry walls.

masonry in combination with application of steel ties and grout injection. Through thicker layers, the amount of reinforcement can be increased and torcrete technology can be applied.

In the case of weak or shallow foundations, it is necessary to strengthen them by widening or deepening them. This is achieved by constructing a reinforced concrete tie-beam along the outside edge of the foundations (fig. 14). New tie-beam should be anchored into the existing foundations by transverse anchors.

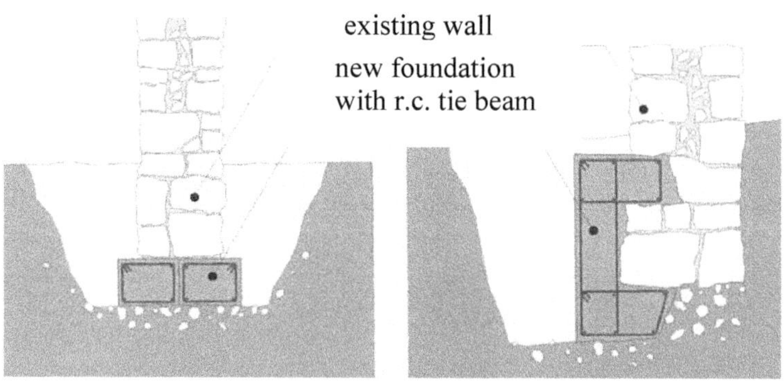

Figure 14: Strengthening of foundations [6].

Only carefully designed combination of strengthening procedures can provide adequate behaviour of the structure during the impact of an earthquake. An example of combining several strengthening procedures on the nave of St Anton the Hermit is shown in fig. 15.

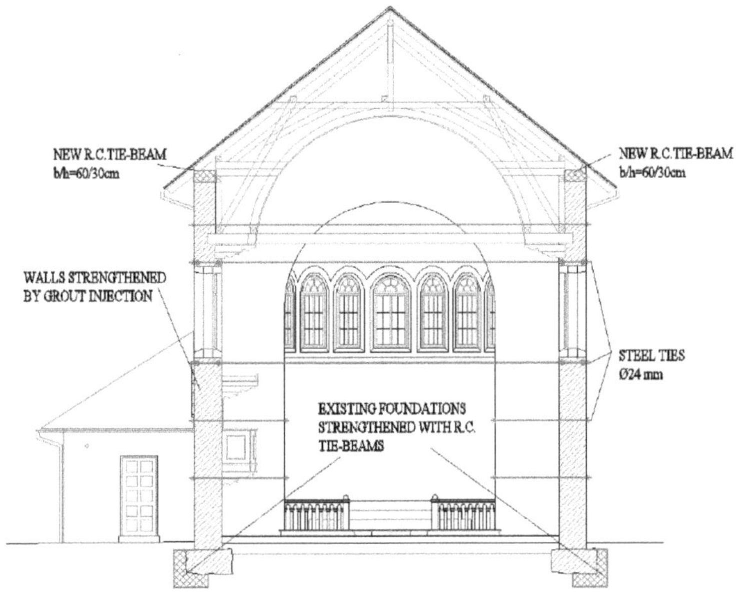

Figure 15: Strengthening procedures on the nave of St Anton the Hermit's church.

6 Conclusion

Some of the basic strengthening procedures applied on churches in the Posočje region are presented. Although churches are mostly of better quality than ordinary buildings, because of their typical design, strengthening procedures are more complex and demanding. Because churches are usually under cultural heritage protection, strengthening must be performed without substantial modifications of the basic structural system. Therefore, adequate seismic resistance is harder to achieve compared to ordinary buildings. In order to achieve optimal results, strengthening procedures should be carefully planned and implemented, based on the results of the preliminary investigations and static and seismic analysis.

References

[1] Naglič, A., Seismic analysis of religious buildings (in Slovenian). Graduation thesis, University of Ljubljana, Faculty of Civil Engineering and Geodesy, Ljubljana, 2011.
[2] Tomaževič, M. Expert evaluation of seismic strengthening of masonry buildings with injected anchors. *ZAG*, Ljubljana, 1998
[3] Gigla, B., Bond strength of injection anchors as supplementary reinforcement inside historic masonry. *13th International Brick and Block Masonry Conference*, Amsterdam, July 4-7, 2004.

[4] Tomaževič, M., *Earthquake-resistant Design of Masonry Buildings.* Imperial College Press: London, 1999.
[5] Corradi, M., Tedeschi, C., Binda, L. & Borri, A., Experimental evaluation of shear and compression strength of masonry before and after reinforcement: deep repointing. *Construction and Building Materials*, **22(4)**, pp. 463-472, 2008.
[6] Deu, Ž., Renewal of residential buildings in the Slovenian countryside (in Slovenian). Kmečki glas, Ljubljana, 2004.

Structural performance evaluation of a historical landmark built in the Taisho period

S. Nagashima[1], M. Mochizuki[1], M. Suzuki[1], T. Ninomiya[1],
R. Tamura[1], T. Nakano[2] & M. Kinoshita[2]
[1]*NTT Facilities, Inc., Japan*
[2]*NTT Facilities Research Institute Inc., Japan*

Abstract

The main building of Yokufu-en has been designated as a Selected Historical Building of the Tokyo Metropolitan Government, built after the 1923 Great Kanto Earthquake in the Taisho period. This building was designed by Yoshikazu Uchida, a famous university professor and his assistant. As a result of the study, the building has sufficient structural strength and the concrete shows sufficient compressive strength. However, as the 3rd story has a high eccentric factor, it is necessary to be reinforced for earthquake resistance. The reinforcement plan was designed for obtaining a prescribed seismic capacity performance in Japan.
Keywords: seismic evaluation, historical-landmark, structural performance evaluation, Yoshikazu Uchida.

1 Introduction

The main building of Yokufu-en (the main building of current Yokufukai, a social welfare corporation), considered in this paper, was a hospital which was built to protect people who were affected by the 1923 Great Kanto Earthquake. It was designed by Yoshikazu Uchida and his assistant Tatuto Toki in 1925 and completed in 1926 [1]. It is a valuable building designated as a Selected Historical Building of the Tokyo Metropolitan Government in 2001. The building is currently used as offices. Structural and seismic evaluations were conducted to check its seismic capacity. This paper reports the details of the technical data used when the building was designed including the Urban Building Law and the structural and seismic evaluation results of the building for the structural performance of this historical landmark designed and constructed in the Taisho period.

2 Outline of the building

Location	: Tokyo
Application	: Offices (original design: hospital rooms, offices)
Designed	: 1925 (Completed: 1926)
Total floor area	: 1,708 m^2
No. of stories	: 3 above ground, 1 below ground, 3-story PH
Structural classification	: Reinforced concrete structure
Structural type	: Rigid-frame structure with seismic resisting walls
Foundation type	: Spread foundation
Design	: Yoshikazu Uchida, Tatuto Toki
Construction	: Obayashi Corporation, Nakamura Koumusyo.

Elevators were installed in this building in 1995 and part of the exterior wall was repaired in 2005. The building has not been affected by severe natural disasters since its completion. Fig. 1 shows a representative elevation and Fig. 2 shows a plan of the representative floor. Its tiled exterior walls have been repaired and part of the surface of the interior wall has been repaired and repainted. Though exposed reinforcing steel was found under part of the eave during appearance and crack check, no obvious cracks, rust on reinforcing steel, or rust fluid were found.

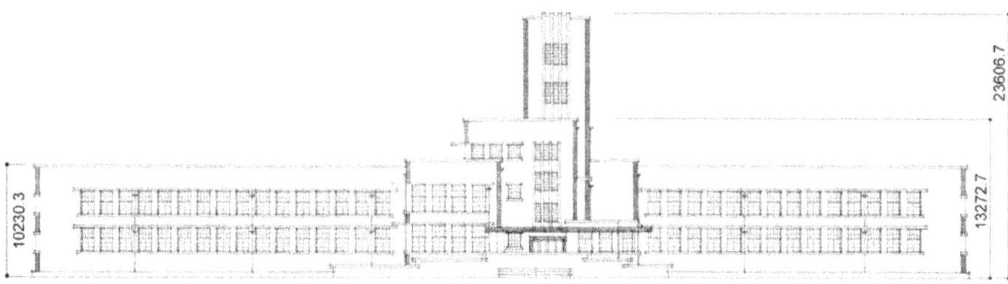

Figure 1: Elevation of the north side.

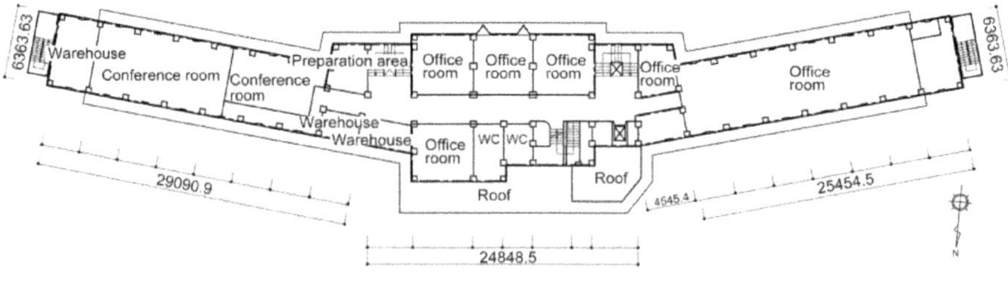

Figure 2: Plan of the representative floor.

3 Structural performance evaluation of concrete and reinforcing steels

3.1 Physical test of concrete

The physical test of concrete was conducted using core samples. Each of the samples (12 in total) was collected from the inside of each room. Table 1 shows where the samples were collected and the results of the compressive strength test. The concrete strength test was conducted in accordance with JIS A 1107 (Japanese Industrial Standard). The corrected compressive strength of the samples is calculated by obtaining the correction factor from the ratio between the height and the average diameter, dividing the largest compressive force by the area and multiplying the value by the correction factor. The average value of all data is 29.9 N/mm^2, and that for above ground is 30.7 N/mm^2. The standard deviations are shown in Table 1. Though the values differ slightly, the strength would be sufficient for Japanese standard for seismic evaluation.

Table 1: Results of concrete compressive strength test.

Code	Floor	Portion	Room name	Corrected compressive strength (N/mm^2)	Standard deviation (N/mm^2)
B1-1	1st basement level	Wall	Dry area	30.1	4.4
B1-2	1st basement level	Wall	Elevator machine room	29.5	
B1-3	1st basement level	Wall	Staircase	22.2	
1-1	1st floor	Wall	Warehouse	24.5	4.6
1-2	1st floor	Wall	Staircase	32.1	
1-3	1st floor	Wall	Conference room	32.8	
2-1	2nd floor	Wall	Warehouse	41.9	8.1
2-2	2nd floor	Wall	Warehouse	32.4	
2-3	2nd floor	Wall	Corridor	25.7	
3-1	3rd floor	Wall	Stairs	31.2	1.9
3-2	3rd floor	Wall	Book room	28.0	
3-3	3rd floor	Wall	Book room	27.9	

3.2 Reinforcing steel strength test

One reinforcing steel sample was collected from the interior wall of the warehouse on the 3rd floor. Table 2 shows the results of the tensile strength test. It also shows the comparison with the provisions of the Urban Building Law revised in 1924 and the current JIS (Japanese Industrial Standard). This building satisfies the standard for the reinforcing steel under the Urban Building Law when it was designed. While the tensile strength is slightly lower than the standard value specified under the current JIS, the yield point and the elongation are SR235 (JIS) level.

Table 2: Results of reinforcing steel tensile strength test.

Standard	Diameter	Yield point (N/mm^2)	Tensile strength (N/mm^2)	Elongation (%)
Measured value	10φ (9.64 mm)	318	367	37
Urban Building Law	10φ	No specification.	353	20 or more
JIS standard SR235	10φ	235 or more	380 to 520	20 or more

4 Consideration of the Urban Building Law when the building was designed and the concrete mix

A bill of architecture was created from the end of the Meiji period to the beginning of the Taisho period. It was finally promulgated as the Urban Building Law in April 1919, and took effect in December 1920 [2]. Yoshikazu Uchida, the designer of the main building of Yokufukai, made a great contribution to the draft of the law. Following the 1923 Great Kanto Earthquake, the provisions of the law relating to structures were revised and the seismic coefficient method was introduced in 1924. It is estimated that the structural design for the main building of Yokufukai designed in 1925 incorporated the contents. The outline of the structural provisions under the Urban Building Law when the building was designed is described below. The Urban Building Law is referred to as "Urban Law" and the Building Standards Law is referred to as "Standards Law" for short. Both SI units and original units are used in comparison. Those for the reinforcing steels are shown in Section 3.2.

4.1 Design live load

Table 3 allows a comparison of design live loads. When the long-term stress is examined, the live load according to the present usage is lower than the design live load defined when the building was designed.

Table 3: Comparison of design live load.

Law	Applications	S For slabs	B For beams	R For rigid-frames		E For earthquakes
				Girders	Column	
Urban Law	Offices, hospital rooms	370 3628	333 3266	296 2903	259 2540	— (not defined)
Standards Law	Offices	300 2900	300 2900	180 1800		80 800

Upper: (kgf/m^2); Lower: (N/m^2).

4.2 Concrete mix

It is estimated that concrete which "passed the provisions for the Portland cement test method set forth in the announcement No. 485 of the Agriculture and

Commerce Ministry" was used from the description of the enforcement regulations for the Urban Law and that the mix was "cement 1: sand 2: gravel or crushed stone: 4" by volume ratio. However, the water-cement ratio was not set forth in the Enforcement regulations for the Urban Law, and therefore, the Architectural Institute of Japan issued the "Concrete and reinforcing steel concrete standard specification" in 1929 and the "Standard for Structural Calculation of Reinforced Concrete Structures" (RC standard) [3] in 1933 in the form of rules and guidelines. However, the relationship between concrete strength and water-cement ratio has already been known academically; Journal of Architecture and Building Science reported that the water volume greatly affects concrete properties [4].

The concrete for the building above ground was placed between March and May. The average temperature in Tokyo in those days was 6 deg C to 20 deg C and the average humidity was 60 to 80%, which were relatively favourable for concrete placement and curing. In addition, the involvement of top-class Japanese engineers at the time also contributed to the strength being increased. The long-term strength of ordinary concrete using normal Portland cement tends to increase in proportion to the 28-day strength under climate conditions close to those mentioned above [5, 6], which could have also improved the strength. Table 4 shows the concrete mix and the allowable stress set forth under the Urban Law in those days.

Table 4: Stress intensity of concrete set forth under the Urban Building Law.

Type	Resistant pressure	Resistant tension	Resistant shear force
Concrete (Material ratio = 1:2:4)	45 4.4	4.5 0.44	4.5 0.44
Concrete (Material ratio = 1:3:6)	30 2.9	3.0 0.29	3.0 0.29

The material ratio is the ratio among cement: sand: gravel (crushed stone). Upper: (kgf/cm^2); Lower: (N/mm^2).

5 Results of neutralization test

5.1 Results of neutralization test

The locations of the neutralization test were the same as those for the concrete compressive strength test. The test results are shown in table 5. The warehouse on the 1st floor (1-1) had the largest depth of 61 mm, but it is approximately 30 to 40 mm as a whole. The entire average is 32.5 mm and the average value above ground is 34.5 mm. Based on this, it is estimated that the neutralization nearly progresses to the position of the reinforcing steel.

Table 5: Results of neutralization test.

Code	Average neutralization depth (mm)	Standard deviation (mm)	Code	Average neutralization depth(mm)	Standard deviation (mm)
B1-1	34.1	5.9	2-1	29.6	6.2
B1-2	15.4	6.8	2-2	29.3	3.0
B1-3	30.3	10.6	2-3	28.4	2.2
1-1	54.6	4.9	3-1	32.1	2.3
1-2	23.6	2.6	3-2	31.7	4.4
1-3	39.1	0.9	3-3	41.7	4.9

5.2 Neutralization curve created from the estimated water-cement ratio

The water-cement ratio estimated from the references in 4.2 and the background of the designer is approximately 0.6. Fig. 3 shows the neutralization curve calculated from the Kishitani formula [7] based on that value and the measured values. The neutralization ratio was set at 1.0. Though there are large variations, the rate of neutralization calculated from the estimated water-cement ratio and the average value show a similar response. As reference data, data of two school buildings which were completed in the same period shown in reference [8] were plotted. (K: Completed in 1925; W: Completed in 1932)

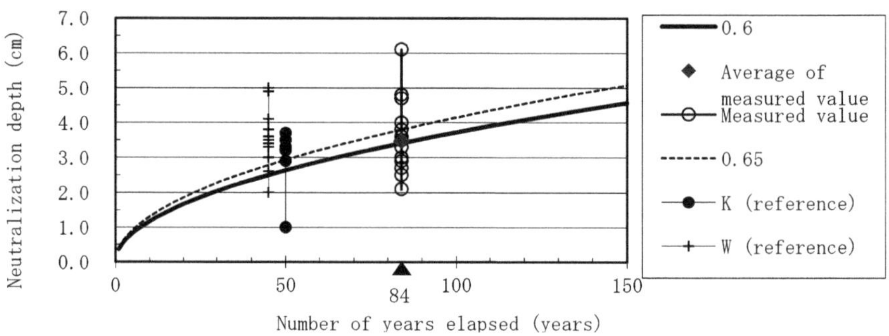

Figure 3: Comparison of measured and calculated neutralization depth.

6 Results of seismic evaluation

6.1 Reinforcing steels

The intervals of bar arrangement were surveyed using the electromagnetic radar system. The bar arrangement was surveyed using 22 columns on the 1st and the 2nd floors. As a result, the authors confirmed that the number of main reinforcement of columns corresponded to the drawing created during the design. The intervals of measurement of the shear reinforcement were 191.2 mm on average, and the largest displacement was within 20 mm. The intervals used for the calculation were set at 200 mm. Fig. 4 to fig. 6 show the drawings used during design. The cross-section chart (fig. 4) only shows the number of windings of shear reinforcement. The detailed structural drawing (fig. 5) shows the number of windings and the intervals. In the drawing, the top and the bottom of the columns are indicated with the number of windings as @0.40 (Shaku) (approx. 121 mm), and the center of the column as @0.60 (Shaku) (approx. 182 mm). However, there was no clear agreement between the cross-section chart (fig. 4) and the number in the detailed structural drawing (fig. 5). For this reason, results of the radar survey were used for column shear reinforcement.

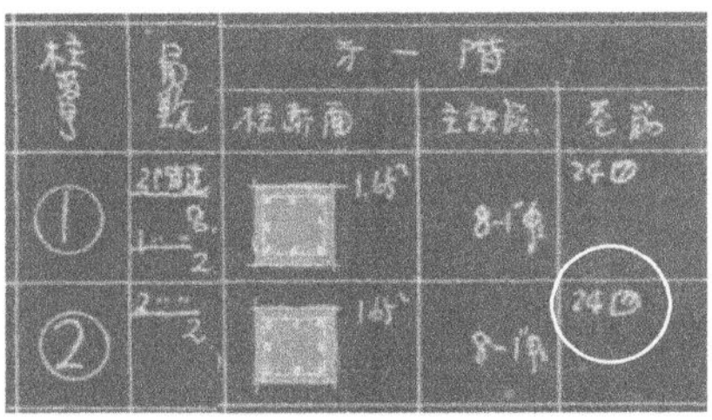

* The drawing of ⍁ in the spiral reinforcement (circled section) indicates the processing of reinforcing steel.

Figure 4: Cross-section chart.

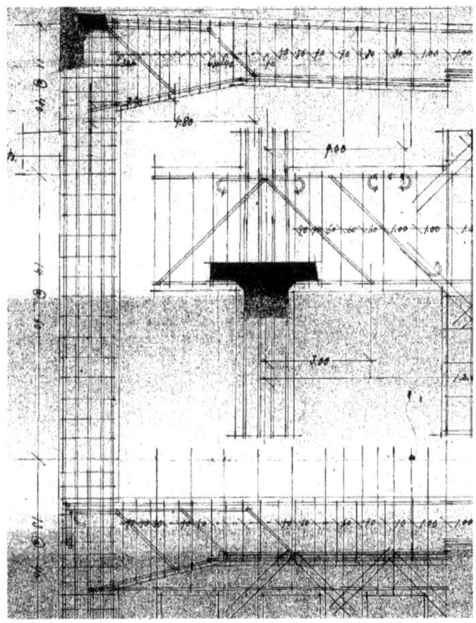

Figure 5: Detailed structural drawing.

For hook shape at the edge of the column shear reinforcement, the figure shows a 135-deg hook (fig. 4). Estimating from the background of architectural engineers, the authors determined that the building was constructed in accordance with the drawing and that the 135-deg hook was used. Several years later, Article 36 of the "Reinforcing steel concrete standard specification" [3] in 1933 specified that the folding of the terminal of joining reinforcement (shear reinforcement) should be 135 deg or more.

Since wall bar arrangement is not found in any drawing, it was determined to be 300 mm pitches with a staggered bar arrangement from the results of actual measurement. The diameter of the wall reinforcement was determined to be 9φ.

For fixing of beam main reinforcement, the detailed beam main reinforcement structure drawing of fig. 6 shows that the top reinforcement exceeds the center of the column and that the bottom reinforcement is folded about 30 cm near the center of the column. Since the smallest column width is approx. 50 cm and the main reinforcement diameter is 3/4 inch (approx. 19 mm), the anchorage length is 29d or more from the column surface.

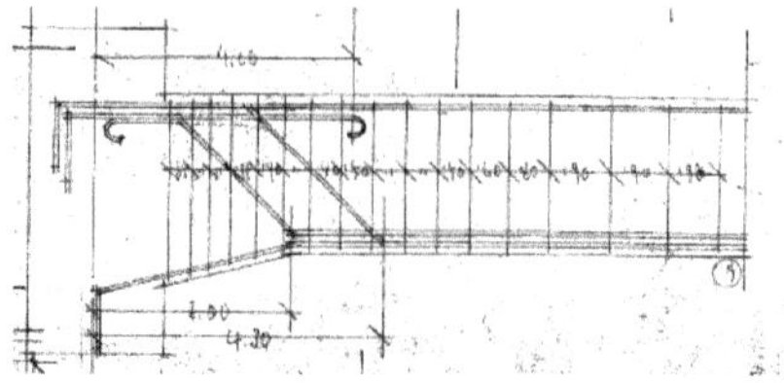

Figure 6: Detailed structural drawing of beam main reinforcement.

Table 6 shows comparison between the structural provisions set forth in the then Urban Building Law (after revised in 1924), the description in design drawings (structural drawings), and the provisions set forth under the present Building Standards Law. The drawings satisfy the structural provisions after their revision in 1924.

Table 6: Comparison between the structural provisions and the drawings.

	Drawing (comparison with the Urban Law)	Urban Law (after revised in 1924)	Standards Law (current rules)
Fixing of reinforcing steel	Beam top: Fixed exceeding the column center Beam bottom: Fixed at the column center.	Fastening of the terminal to other structures or fixing by folding the terminal.	Designation of the terminal position to start folding for fixing Terminal of round bar, with hook
Joint length of main reinforcement	Beams satisfy the standards from the structural drawing (Diameter of beam main reinforcement: 19φ) Columns were checked from the photo taken during construction.	25D or more of the main reinforcement diameter	Tensile reinforcement of the beam for fixing the column: 40D or more Location with small tension: 25D or more
Beam shear reinforcement	The largest interval is 1 Shaku (303 mm) with the beam center depth of 500 mm, which satisfies the Urban Law.	Placement of binding reinforcement Intervals of binding reinforcement: 2/3D or below	Beam: Stirrup reinforcement 3/4D or below
Beam reinforcing steel	Double reinforcement	Placement of double reinforcement	Placement of double reinforcement
Column structure	8 to 16 columns	4 or more main reinforcement	4 or more main reinforcement
	The reinforcing steel area of all the columns covers 1/80 or more of the column area. (Column main reinforcement diameter: 25φ)	1/80 (1.25%) or more of the concrete cross-section area	0.8% or more of the concrete cross-section
	Indication of the number of windings (Average of measured value: 191.2 mm)	Intervals of binding reinforcement are 1 Shaku (303 mm) or below and less than 15 times the main reinforcement diameter	10 cm for column top/base (15 cm for others) and less than 15 times the thin main reinforcement diameter Hoop reinforcement ratio: 0.2% or more
	All the columns satisfy the standards.	Column small diameter: 1/15 or more of the major spans	Column small diameter: 1/15 or more of the major spans

6.2 Results of seismic evaluation

Based on the results of the field survey and the material test, the Standard for Seismic Evaluation of Existing Reinforced Concrete Buildings, 2001 (The Japan Building Disaster Prevention Association) were used to create a seismic evaluation and reinforcement plan. From the average value above ground obtained from the results of the concrete compressive strength test (see 3.1), the concrete strength for calculation was set to Fc28. For reinforcing steels, though the tensile strength is slightly lower than the JIS standard, the yield point and the elongation are satisfied. Therefore, they were set to SR235.

The entire building was evaluated as a six-layer building because the proportion of the 4th or higher floors, which are the penthouse, is large when compared with the 3rd floor in terms of both the area and the weight. The building weight by unit area is larger than that of ordinary buildings (table 7); this is caused by relatively small spans, a high proportion of exterior walls due to the wing-shaped form, the exterior walls finish with brick-like scratch tiles and other factors.

Table 7: Area and weight of each floor.

Floor	Floor area of each floor Af (m^2)	Total floor area Σ Af (m^2)	Total weight Σ W (kN)	Unit weight w_2 (kN/m^2)	Ai
6 (PH3)	26.5	26.5	467	17.7	3.16
5 (PH2)	26.5	52.9	1016	19.2	2.45
4 (PH1)	26.5	79.4	1623	20.5	2.14
3	114.4	193.8	3460	17.9	1.75
2	739.5	933.3	13145	14.1	1.25
1	785.4	1718.7	24692	14.4	1.00

PH: Penthouse.

Fig. 7 to fig. 9 show framing plans and framing elevations of representative sections. Most of the wing-shaped sections (AX1 - AX8, CX1 - CX8) have a rigid-frame structure, but seismic resisting walls are provided to the terminal in both *X* and *Y* directions. The core area at the center of the building has frames with seismic resisting walls around the corridor and room boundaries.

Figure 7: 2nd floor framing plan (EW indicates seismic resisting walls).

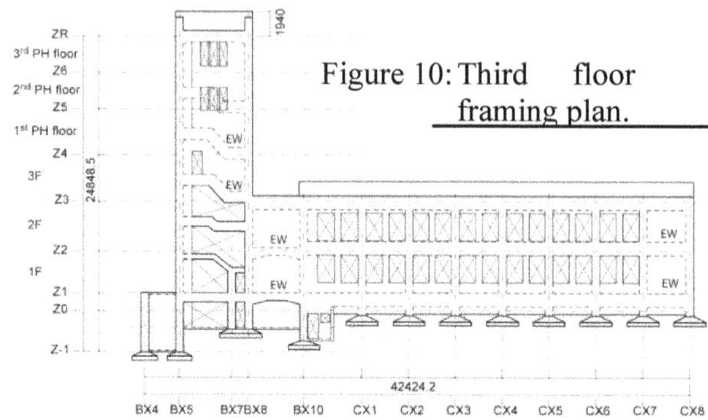

Figure 10: Third floor framing plan.

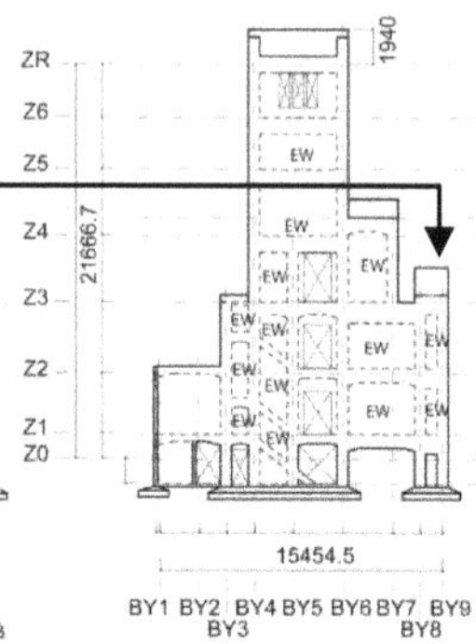

Figure 8: BY4-CY1 framing elevation.

Figure 9: BX5 framing elevation.

The aging index was set at 0.927 based on the results of the field survey. No structural cracks were found during the appearance check. From the progress of neutralization, the neutralization range was set to 1/3 or more of the total members. Though no rust fluid or expansion cracks were found in the corrosion of the reinforcing steels, they were set below 1/9 of the members taking neutralization into consideration.

The seismic evaluation results show that all the floors except the 3rd satisfy the Is value: 0.6 or more and $C_{TU} \bullet S_D$: 0.3 or more (table 8). Spans are small, and the entire building is relatively rigid. Many vertical members have the F-value of 1.0, indicating that the building is resistant to earthquakes with its strong structure. Since the design seismic intensity when the building was designed was 0.1 common to every floor, and was not specified for the penthouse, which was set according to necessity, higher floors see more difficulty compared with the

Table 8: Seismic evaluation results (present condition).

	Floor	C	F	E_0	S_D	I_S	C_{TU}, S_D	Evaluation
X direction	6 (PH3)	6.03	1.00	1.61	0.95	1.42	1.53	OK
	5 (PH2)	5.10	1.00	1.79	0.95	1.57	1.70	OK
	4 (PH1)	2.33	1.00	0.95	0.95	0.83	0.90	OK
	3	1.59	1.00	0.81	0.63	0.47	0.51	NG
	2	2.41	1.00	1.83	0.95	1.61	1.74	OK
	1	1.58	1.00	1.58	0.95	1.39	1.50	OK
	Floor	C	F	E_0	S_D	I_S	C_{TU}, S_D	Evaluation
Y direction	6 (PH3)	5.18	1.00	1.38	0.95	1.22	1.31	OK
	5 (PH2)	5.56	1.00	1.95	0.95	1.72	1.85	OK
	4 (PH1)	3.24	1.00	1.32	0.95	1.16	1.25	OK
	3	1.76	1.00	0.89	0.69	0.57	0.62	NG
	2	1.81	0.80	1.10	0.95	0.96	1.30	OK
	1	1.19	0.80	0.95	0.95	0.83	1.12	OK

The codes in the table are based on the references [9].
Aging index T: 0.927, Criteria: Iso = 0.60 < Is, $C_{TU} \bullet S_D > 0.30$.

present seismic evaluation standard. In spite of this, the 3rd floor and the penthouse are strong due to their small area and many walls. However, the position of the penthouse (4th to 6th floors) has a planar eccentricity against the plane arrangement of the 3rd floor. Therefore, the Is value fell below the standard value (= 0.6) due to the eccentricity ratio of the 3rd floor of 0.359 in the *X* direction; 0.241 in the *Y* direction.

7 Seismic reinforcement plan and seismic capacity

From the results described in Section 6, the reinforcement plan was made to increase the bearing force in the *X* direction on the 3rd floor, and solve the eccentricity ratio caused by the penthouse in the *Y* direction. To solve the eccentricity ratio, the seismic reinforcement plan took into account that the building is a historical landmark, and increased concrete of the interior walls to increase the strength and reduce the eccentricity ratio, not providing slits to the walls so that the appearance of the building would not be affected. Fig. 10 shows the details of the reinforcement plan.

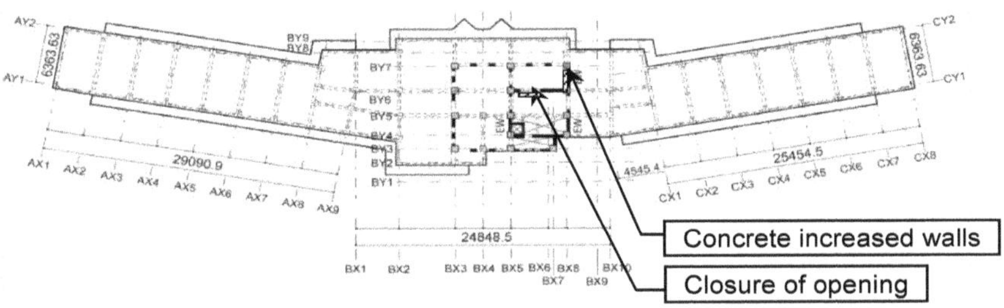

Figure 10: Seismic reinforcement location chart (3rd floor framing plan).

Table 9: Seismic evaluation results (after reinforcement).

	Floor	C	F	E_0	S_D	I_S	C_{TU}, S_D	Evaluation
X direction	6 (PH3)	6.03	1.00	1.61	0.95	1.42	1.53	OK
	5 (PH2)	5.10	1.00	1.79	0.95	1.57	1.70	OK
	4 (PH1)	2.33	1.00	0.95	0.95	0.83	0.90	OK
	3	2.18	1.00	1.11	0.63	0.65	0.70	OK
	2	2.41	1.00	1.82	0.95	1.61	1.73	OK
	1	1.58	1.00	1.58	0.95	1.39	1.50	OK
	Floor	C	F	E_0	S_D	I_S	C_{TU}, S_D	Evaluation
Y direction	6 (PH3)	5.18	1.00	1.38	0.95	1.22	1.31	OK
	5 (PH2)	5.56	1.00	1.95	0.95	1.72	1.85	OK
	4 (PH1)	3.24	1.00	1.32	0.95	1.16	1.25	OK
	3	1.77	1.00	0.90	0.78	0.65	0.70	OK
	2	1.81	0.80	1.10	0.95	0.96	1.30	OK
	1	1.19	0.80	0.94	0.95	0.83	1.12	OK

The codes, the aging indices, and the criteria in the table are the same as those of Table 8.

The plan increased the bearing force in the *X* direction and improved mainly the S_D index in the *Y* direction. Consequently, Is and $C_{TU} \cdot S_D$ satisfied the criteria values on all the floors. Table 9 shows the calculation results after the reinforcement plan was implemented.

8 Conclusion

This paper reported the seismic capacity of a building for disaster reconstruction which was designed and constructed soon after the 1923 Great Kanto Earthquake at the end of Taisho period as an example.

Yoshikazu Uchida, a concrete architectural engineer, made a great contribution to this building. The building shows his design purpose for constructing a building which is highly rigid and can endure a great earthquake in response to the damages caused the 1923 Great Kanto Earthquake. Since the concrete is strong, it is estimated that the building was constructed carefully. From the drawings used during the design and photos during construction, the authors learned exemplary bar arrangement methods before the standards including the Architectural Institute of Japan were created.

References

[1] Outline of the construction work of Yokufu-en; Yokufukai, "Materials on Yoshikazu Uchida" stored at Tokyo Metropolitan Archives, April 1927

[2] Urban Building Law, 1920; revised in 1924

[3] Reinforcing steel concrete standard specification; Standard for Structural Calculation of Reinforced Concrete Structures, Architectural Institute of Japan, 1933

[4] Doi, M., Second report on concrete strength test (lecture on comprehensive construction material survey. *Journal of Architecture and Building Science*, 1921.

[5] Takahashi, N. & Kubota, N., Study on strength control of structural concrete, Part 16: Actual condition and examination of core strength of each member in long-age concrete. *Summaries of Technical Papers of Annual Meeting Architectural Institute of Japan*, 1980.

[6] Tomosawa, M., Ko, A., Tanaka, Y., An experiment on strength development of concrete under various temperature and curing conditions. *Summaries of Technical Papers of Annual Meeting Architectural Institute of Japan*, 1985.

[7] Izumi, I., Maeda, T. & Kita, T., Concrete structure durability series: neutralization, Gihodo Shuppan, 1986.

[8] Kimura, K., *Concrete Strength and Durability of RC Constructions*, Kajima Institute Publishing, 1981.

[9] Seismic evaluation and seismic reinforcement design manual for existing constructions for Year 2003, Japan Association for Building Research Promotion, 2003.

Seismic analysis and retrofit of a historic masonry building

C. C. Spyrakos[1], P. Touliatos[2], D. Patsilivas[1], G. Pelekis[1], A. Xampesis[1] & Ch. A. Maniatakis[1]
[1]*Department of Civil Engineering, National Technical University of Athens, Greece*
[2]*Department of Architectural Engineering, National Technical University of Athens, Greece*

Abstract

This study presents the seismic analysis of a three-storey unreinforced masonry building that was heavily damaged during the Aigion 1995 earthquake. The building is located a few kilometers from the fault, thus subjected to "near field" ground motions. Seismic analysis of the building, using an elaborate finite element model with properties extracted from material testing of representative structural members and excitation from records obtained from a nearby station, revealed the mechanism of partial failure of the top floor as well as the weaknesses of the original building design and construction. Alternative retrofit and strengthening techniques that can be implemented while maintaining the architectural and aesthetical appearance of the historic building are presented.
Keywords: seismic analysis, retrofit of masonry building, near-field seismic motion.

1 Introduction

The city of Aegio has suffered several strong earthquakes through the years. The most recent powerful earthquake which caused human losses occurred on June 15, 1995. It was a 6.4 Ms shallow depth earthquake with an epicentral distance of about 18 km southwest of Aegio that caused the death of 26 people, massive, and in many cases, irreparable damages to both modern and older (masonry) buildings [1]. In 1997 a weaker earthquake caused additional damage on already deteriorated buildings.

The three story Panagiotopoulos' mansion is located at the city centre of Aegio (fig. 1). It has a basement of similar dimensions with the other floors. Its plan is rectangular with dimensions 18.68 m by 14.65 m and height of floors of about 4.5 m. The timber roof is rectangular with a Byzantine tile coating. The arrangement of openings on the front views of the building is well proportioned to a vertical axis, and the same on the first and second floor.

The external walls are made of sack masonry with thickness that is reduced with height. They are about 95cm thick at the ground floor and 86cm thick at the second floor. The internal walls are 80cm thick at the ground floor, 75cm at the first and 68cm at the second floor. The walls are three-layered with exterior sides constructed with large boulder (stones) connected with mortar and smaller stones with thicker mortar placed between them. In the whole building, there are quite a few closets incorporated in the masonry walls that reach the height of interior openings. The openings of the external walls are formed at the upper part with shallow arches made of bricks. About 18% of the external walls are openings to the northern, southern and eastern view. To the western view, about 10% are openings.

In the interior of the building there are two walls made of stone masonry. They are parallel to the narrow sides of the building and continue all the way to the roof. Communication among the floors is made through marble stairwell at the southern side of the building, which is supported by stonework, arches and horizontal beams.

The light internal partition walls on the first and second floor of the building are made of bricks with a thickness of 17cm to 18cm supported on either powerful masonry walls or wooden floor beams. At various heights, along each of the interior walls there are two iron flat-shaped steel bars at about 3cm distance between them that augment and increase the bending and shear stiffness of the walls.

The first and second story floors are made of timber. The floors of the ground floor are either timber or stone masonry domes located at the central part of the building where the wall openings of the basement are relatively small. It should be noted that the arrangement of the planks on the first and second floor is made in such a way that the orientation of wooden beams distributes the weight of the floors evenly along all the load carrying walls, so in case of an earthquake, the inertia forces are shared by all vertical walls. At the ground floor steel or timber bars of rectangular profile are anchored with metal plates to the transverse walls, in order to avert detachment of the interior walls at the corners (fig. 2).

At the level of the roof base along the stonework there is also a crossbeam perimetrically to the interior side of the external masonry walls of the building. The cross-beam is attached to the stonework through metal sheets placed at about 1.5–2 m. The cross-beam plays a significant role in the static and seismic performance of the building since it: i) distributes the concentrated loads of the roof to the stone work averting local cracks due to compression; ii) contributes to the development of rigid diaphragm behavior at the floor levels and iii) functions as a fastening of the stonework augmenting the "integral" behavior of the building for seismic loads.

Figure 1: Southern view of building.

Figure 2: Horizontal ties at corners.

The building structural system is called a system of "poles on pillars" and presents two primary advantages: i) a multiple potential to redistribute the loads when some of its component fails and ii) it develops similar behavior along its two principal axes [2].

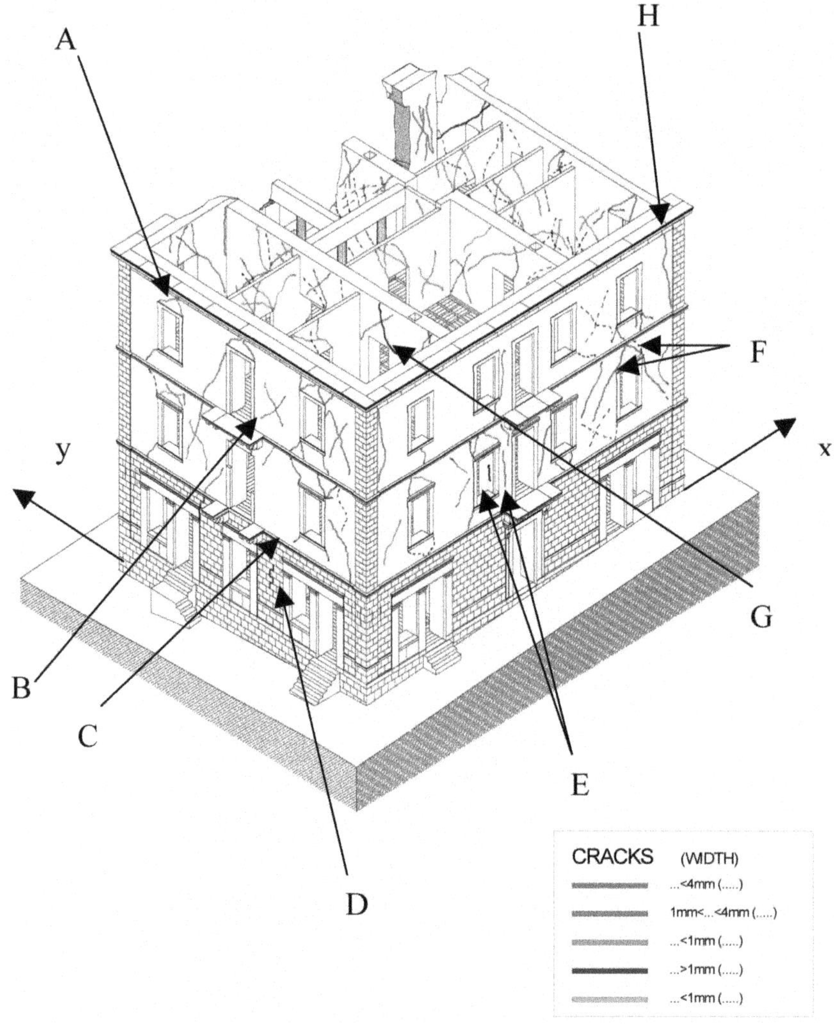

Figure 3: Damage distribution from on-site inspection.

2 Description of damages

The present condition of the building is disappointing because of the extent of the damages caused by the 1995 and 1997 earthquakes. What has played an important role to the damages is, on the one hand, the insufficient maintenance prior to the earthquakes and on the other, not taking any necessary protection measures after the 1995 earthquake.

Fig. 1 shows a large part of the external wall on the second floor that has collapsed during the 1995 earthquake. The southwestern corner of the building on the second floor has also collapsed. A qualitative interpretation of the representative damages shown in fig. 3 is presented below:

A. A crack of parabolic form over a window opening caused by inadequate bending strength. After the collapse of the southern external wall, the rest of the stone masonry carries even higher compressive loads.
B. "Shear" cracks forming an "X" caused by the alternating direction of seismic action. The inclination and form of the cracks depends on the vertical compressive loads carried by the walls, the position and size of the openings [3].
C. Cracks round the openings developed by the seismic forces parallel to the wall.
D. A crack located at the level of the arch at the ground floor. The crack is caused by either horizontal thrust acting on the masonry wall resulting in bending stresses and the co-existence of the intense shear stresses in this area.
E. Vertical cracks in the panels between windows caused by intense compression.
F. Diagonal crack in a wall with reduced thickness to form the chimney.
G. Cracks caused by insufficient connection (lack of a cross beam) of the floor beams at the first and second floor.
H. Vertical cracks at the top of the intersections of transverse walls. There is a tendency of separation between the walls which denotes insufficient strength at their intersections.

2.1 Analysis for seismic loads and correlation with observed damages

The building walls were modeled and analyzed using finite elements [4]. Linear elastic behavior was assumed using three and four node thick shell elements, a process usually adopted for the analysis of similar structural systems, e.g. [3, 5]. Beam elements were used to model the timber parts, such as the floors. The response spectrum analysis that was carried out included a sufficient number of modes, as specified by the Code [6].

Fig. 4 shows the 8th mode of the building with $T = 0.17$ s which is the first most significant mode along the X direction with the largest participation factor (–39.65) and modal mass (62.82%). The 9th mode is the first most significant mode along the Y direction with a participation factor of 31.02 and modal mass of 38.46%. The 9th mode exhibits an intense out of plane bending deformation of the flexible partition walls. It should be noted that the most significant modes in each direction are translational with out-of-plane bending of the exterior walls combined with deformation of the interior walls to maintain compatibility [7].

The inplane deformations of the walls corresponding to the first nine modes are insignificant. The second most significant mode is the 10th mode in the Y direction and exhibits in plane deformations of opposite side walls. The transverse internal walls and especially the most flexible partitions follow the response of the external walls forced by displacement compatibility at their intersections resulting in bending deformations.

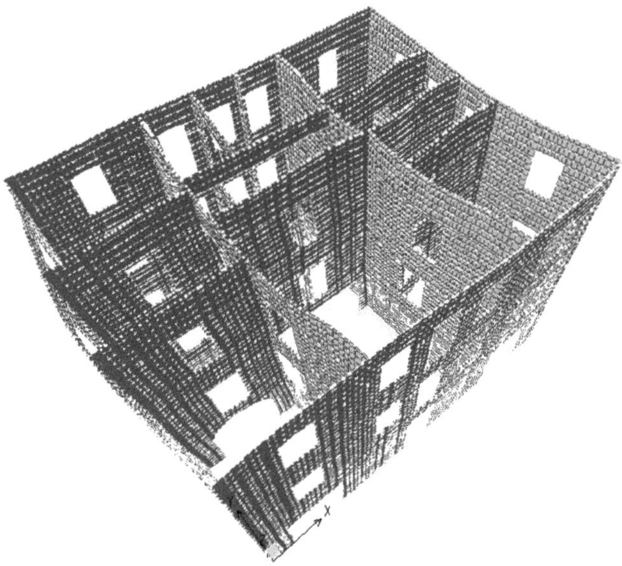

Figure 4: 8th mode shape.

The building was subjected to the response spectra of the Aegio 1975 earthquake, figs. 5 and 6. It must be emphasized that due to the orientation of the building along the direction of the coastline (which is also the direction of the earthquake fault); the spectra were applied along the *X* direction of the building, whereas the transverse spectrum was applied along the *Y* direction. This selection was based on the fact that the *X* direction of the building is parallel to the coastline. The spectra correspond to horizontal ground acceleration recorded a few kilometres from the mansion. The near-field ground motions are characterized by pulses with a maximum acceleration of 0.54g, a long period of about 0.5 s along the transverse direction to the fault and velocity of about 52 cm/s [8].

Masonry structures, such as the Panagiotopoulos building in Aegio have a small out of plane bending stiffness. The Aegio earthquake caused mostly bending out of level response and, as a result, tension prevailed over the construction with detrimental effects. The dynamic analysis showed that the building response was primarily affected by the translational modes along the *X* and *Y*-axes and that for the largest part of the building seismic demand surpassed the strength of the masonry.

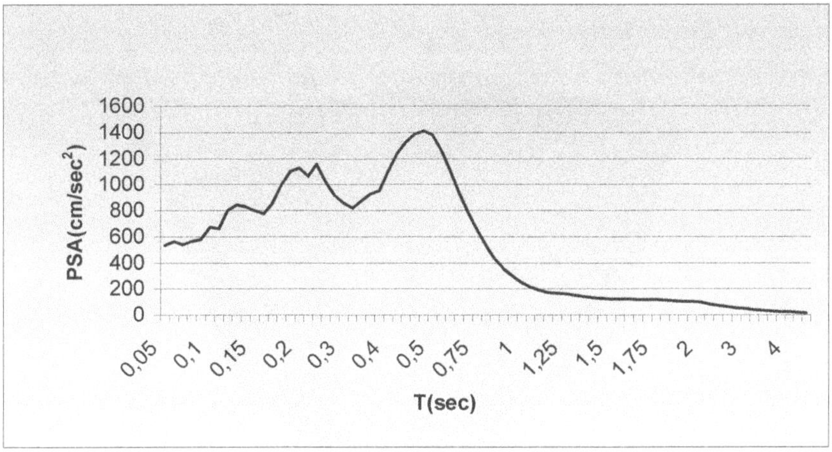

Figure 5: Spectral acceleration in the *Y* direction.

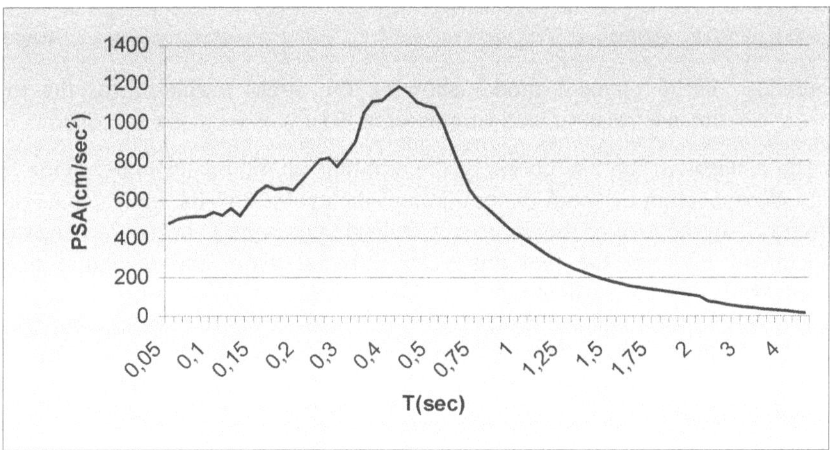

Figure 6: Spectral acceleration in the *X* direction.

Use of a linear elastic model does not allow redistribution of stresses once a part of the structure has failed. However, use of this model demonstrates with remarkable accuracy the building parts that were either the first to fail or suffered considerable damage. In the area of the staircase at the 2nd floor level, indicated with (A) in fig. 7, the largest tensile stresses develop; this implies that this area is one of the first to fail during the earthquake. Out of plane failure is also favored with the internal partition walls. Moreover, after the collapse of the upper part of the wall in the stairwell, an increase of the tensiles stresses on the adjoining masonry is anticipated resulting in its collapse at a second phase, a fact demonstrated by the collapsed area in the southern view of the building shown in fig. 1 [7].

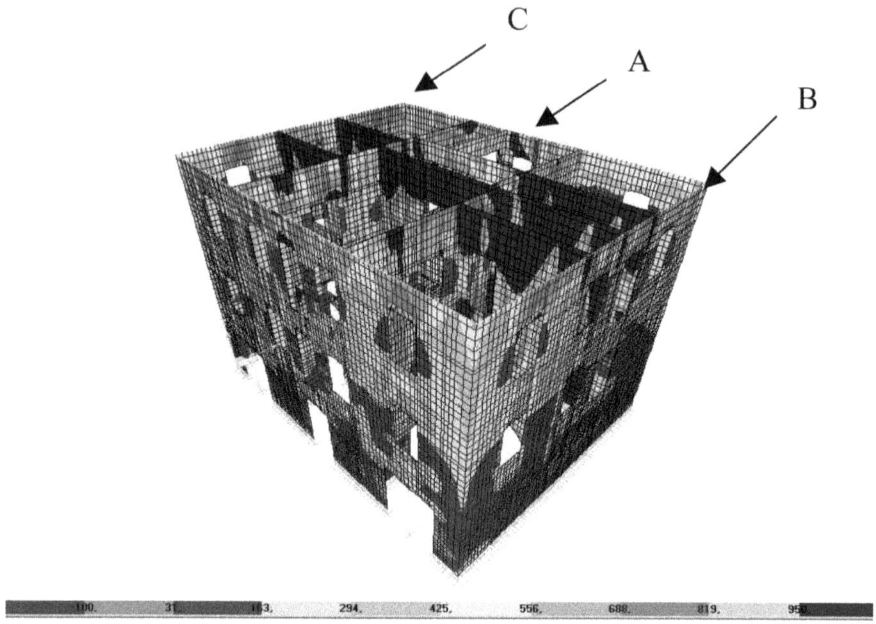

Figure 7: Finite element model showing the stress variation for the most intense seismic load combination [7].

The collapse of the SW corner of the building should be attributed to the lack of good connection between the two transverse external walls, indicated as point B in fig. 7. In the area of the SE corner, indicated as point C in fig. 7, there was a detachment between the two transverse external walls and deviation of the eastern wall from the vertical.

The Aegio earthquake did not cause overloading for the timber roof of the building. Both the tensile and the compressive axial stresses in the wooden truss members of the roof were less than one-fifth of their tensile and compressive strength [7].

3 Proposed interventions and conclusions

The evaluation of the present condition of the structure in conjunction with the results of dynamic analysis leads to the recommendation of the following interventions [7, 9, 10]:

1. Construction of a concrete kerb to create an overall "box" connection at the top of the masonry walls; connection of the kerb to the masonry walls with steel bars and dowels to the timber roof.
2. Enhancing most of diaphragm behavior by strengthening the connection of the timber floor to the masonry walls and replacement of the timber elements, where it is necessary to strengthen diaphragm function.
3. Demolition and reconstruction of the inclined walls.
4. Reconstruction of the damaged SW corner and strengthening of the junctions of the transverse walls.

5. Reconstruction of the collapsed area and strengthening of the masonry walls at the stairwell.
6. Repair and homogenization of the stonework mass.
7. Strengthening of the reduced mass of the stonework in the areas of fireplaces.
8. Repair and strengthening of the stairwell.
9. Reconstruction of the roof by reusing old intact materials.
10. Use of pre-stressed tie-bars to connect intersecting walls.

A detailed discussion and presentation of an extensive repair and strengthening scheme is presented in [7]. A thorough evaluation of alternative schemes and corresponding dynamic analysis of the building showed that the construction of a concrete kerb at the roof level combined with enhancement of diaphragm action could lead to a system capable to suffer minor damages even to the Aegio near-field ground motion.

References

[1] Papazachos, B. & Papazachou, C., *The earthquakes in Greece*, Ziti Editions: Thessaloniki, 1997.

[2] Touliatos, P., *Seismic behavior of traditional and historic buildings in Greece*, Report to O.A.S.P., 2001.

[3] Groci, G., *The Conservation and Structural Restoration of Architectural Heritage, Advances in Architectural Series*, Computational Mechanics Publications: Southampton, UK, 1998.

[4] Spyrakos, C.C., *Finite Element Modeling in Engineering Practice*, Algor Publishing Division: Pittsburgh, P.A., 1995.

[5] Karantoni, F.V. & Fardir, M.N., Computed vs. observed seismic response and damage of masonry buildings. *Journal of Structures Engineering ASCE*, **118(7)**, pp. 1804-1821, 1992.

[6] Earthquake Planning and Protection Organization, Greek Aseismic Code EAK2000. Athens, Greece, 2000.

[7] Patsalivas, D., Pelekis, G. & Hambesis, A., *Structural Renovation of the Panagiotopoulos Building in Aegion*, Diploma thesis, N.T.U.A, 2004 (in Greek).

[8] Spyrakos, C.C., Maniatakis, C. A. & Taflabas, J., Critical evaluation of near field seismic records in Greece. *ERES V*, eds. C. A. Brebbia, D. E. Beskos, G. D. Manolis & C. C. Spyrakos, pp. 53-62, 2005.

[9] Tomazevic, M., *Earthquake Resistant Design of Masonry Buildings*, Imperial College Press: London 2001.

[10] Spyrakos, C.C., *Strengthening of Structures for Seismic Loads*, T.E.E., Athens, Greece, 2004 (in Greek).

Non destructive testing techniques applied to the masonry and timber structures of the Crocifisso Church in Noto

L. Binda, L. Cantini, P. Condoleo & A. Saisi
DIS – Dept. of Structural Engineering, Politecnico of Milan, Italy

Abstract

After the partial collapse of the Cathedral of Noto in Sicily, damaged by the 1990 earthquake, which hit the Eastern part of the Italian island, an extensive investigation carried out on site and in laboratory allowed the detection of a pre-existing damage of the pillars due to long term behaviour of the material under compressive stresses. The investigation extended to other churches and buildings showed cases of progressive damage, which can in some unforseen time, bring the structures to sudden collapse. This paper presents and discusses the preliminary results of the research carried out by the authors on the Crocifisso Church in Noto, built in the same period and with material similar to those of the Cathedral. The investigation programme was planned with the aim to evaluate the state of damage of the structure and to design the preservation and restoration actions.
Keywords: diagnosis, on site test, long-term behaviour.

1 Introduction

The dramatic collapse of the central nave and of part of the dome of the Noto Cathedral (fig. 1) damaged by the 1990 earthquake, took place only six years later, in 1996 [1]. The accurate on site and laboratory investigation confirmed that the earthquake was only worsening the situation of the materials and of the structure already highly damaged [2, 3]. The removal of the plaster (which was made in the sixties) from the remaining pillars of the central nave revealed the pre-existence of a diffused distribution of vertical cracks and spalling of the corners in the external leaf made of limestone (calcarenite) regular blocks (fig. 2). In fact, these cracks had been filled with the gypsum mortars used for plastering [2]. The continuous damage of the masonry under high compressive stresses is considered to bring about collapse in the long term. Other cases were found to be characterised by the same phenomenon as the collapse of the Civic Tower of Pavia in 1989 [4], of the Goch bell tower in 1992 and most recently the Meldert Bell-tower and the Maagdentoren Tower at Zichem (Belgium, 2006).

In fact, it has been demonstrated that the behaviour of masonry under the action of persistent loads can evolve in a relatively long time until collapse; this can happen under lower stress values than those corresponding to the nominal material strength obtained by a standard monotonic compression test. The phenomenon can start for some masonry type at 45-50% of the nominal strength value [5]. The available experimental data collected so far tend to show an evident increase of lateral deformations progressing through time caused by the development of typical vertical cracks due to compressive stresses. This dilation phenomenon, an apparent increase in volume, can lead to collapse due to crack propagation. The effects can also be coupled with synergetic stresses caused by cyclic wind action and temperature variation. Additional minor shocks, like storms, low intensity earthquakes, etc., may contribute to an increase in damage. The damage can easily develop when the material used for the construction is rather weak (weak bricks and mortars with irregular joints, soft stones) or the technique of construction is such that the internal core of the masonry can settle and deform more than the external leaf. In all these cases the damage can start early, even a few years after the construction or after some partial reconstruction or even after repair of the masonry and continue very slowly for decades, until a sudden collapse happens.

Figure 1: The Cathedral of Noto after the 1996 collapse.

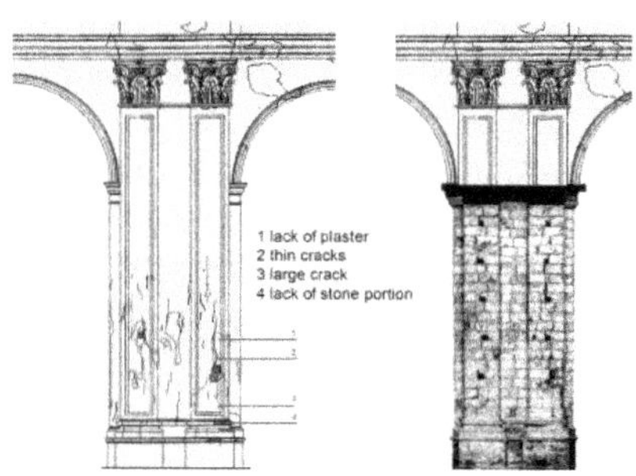

Figure 2: The damage survey of a Noto Cathedral pillar.

In the case of the Noto Cathedral, the internal rubble masonry of the pillars was found to be very weak [2]; its early higher deformation caused in the long range a high state of damage in the external stone leaf (fig. 2). A further weakness was found in the materials used for the external leaf of the collapsed pillars. In fact while for the internal rubble filling the thick mortar joints were made by hydrated lime and silt weak aggregates alternating with rather irregular courses of round river stones and there was a lack of connection between the internal rubble and external leaves, these were made of regularly cut blocks of a very weak travertine. These characteristics did certainly contribute to create weak load-bearing elements, which became in the long range highly stressed especially in the external stiffer leaves and the situation evolved toward increasing damage. The materials used and the construction technique of the load-bearing elements were both typical of the area; in fact the presence in the eastern part of Sicily of several quarries of limestone (also called "Noto stone") which could be easily carved, made this material very popular for the construction of churches, noble palaces and other monuments. In addition, the mortar aggregates were very frequently obtained from the same stones and quarries [2].

Several other churches and monuments were subjected to investigation to detect the extension of the phenomenon [6]. Among them, the Crocifisso Church in Noto was studied in order to evaluate its state of damage after the 1990 earthquake. The materials and technique of construction used for this church are very similar to the ones of the Noto Cathedral.

The investigation carried out on the structure of the Crocifisso church and described here, showed a diffused damage in the load-bearing elements, especially in the pillars of the nave, similar for materials and technique of construction to the ones of the collapsed Cathedral. Therefore, a quick repair and preventive intervention was necessary for the weakest parts.

2 The Crocifisso Church: damage description

After the earthquake of 1693 that destroyed the ancient city of Noto and the decision to rebuild the new Noto, the Crocifisso Church, placed in the highest part of the new town of Noto was built using the local limestone and finished in 1715. The façade (fig. 3) and the interior show harmonious proportions, with rather slender pillars. A sequel of small domes covers the aisles; the central nave, instead, shows a barrel vault and a large dome. Above the barrel vault, a timber truss is present.

The church was built using the local limestone, the same as the one of the Cathedral and was finished in a relatively short time after the earthquake; also here, due to the high number of buildings in the new town, the construction was made probably with a poor workmanship. In fig. 4, the Church vertical section is shown.

Figure 3: View of the façade of the Crocifisso Church.

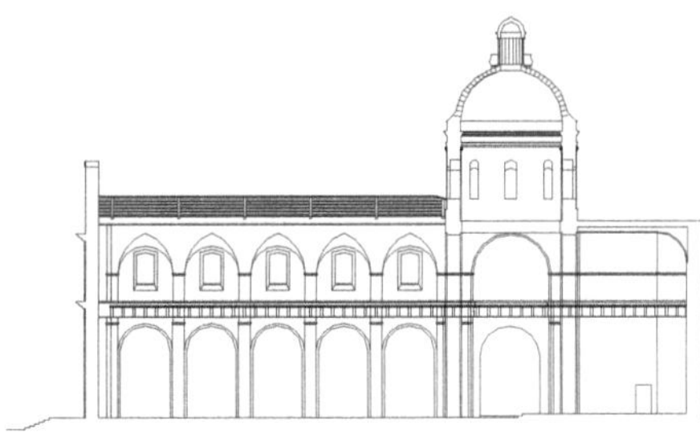

Figure 4: Longitudinal section of the Crocifisso Church.

The church was damaged by the 1990 earthquake, especially the transept and domes of the lateral aisles and the vaults, which were therefore supported by provisional structures. The pillars, all covered with plaster and stucco, did not show any apparent damage.

In contrast, during the first site inspection phase of the diagnosis an alarming state of damage was found in the pillars. The plaster, a thin one (4 to 5 mm thick) made by hydrated lime based mortar, was in fact hiding a pattern of vertical cracks in the stones continuing along the mortar joints. The reason was that the plaster was detached from the pillar surface and hence was hiding the real damage. After completely removing the plaster (made in the sixties), it was possible to perform a crack pattern survey of the pillars.

The results of the experimental investigations are reported together with the suggested repair intervention in the following sections.

2.1 Damage after the earthquake

As mentioned above, the Church was damaged by the 1990 earthquake, especially the transept, the domes of the aisles and the vaults, while the pillars, all covered with plaster and stuccoes, did not show any apparent damage. A simple calculation of the state of stress due to uniformly distributed dead load gave 0.72 to 0.81 N/mm^2. The maximum compression value caused by horizontal forces plus dead load was calculated as 1.20 to 1.59 N/mm^2; apparently, the state of stress did not seem to be too high.

3 Investigation program

In view of a strengthening and repair intervention, an investigation program was planned. The program included the survey of the foundation typology by coring and boroscopy and the study of the damage of the masonry structures by flat jacks and sonic tests, mortar and stone characterisation [7].

A survey was also performed on the timber truss of the roof, which had a very large dimension (almost 10 m) and appeared to have some damage. Geometrical survey, penetrometry tests by resistograph and sonic tests were carried out on these structures in order to detect their state. Fig. 5 shows the church plan with the localisation of the tests on the masonry.

3.1 Masonry structures

3.1.1 Crack pattern survey

As mentioned above, an alarming state of damage was found in the pillars, after locally removing the plaster. The plaster, in fact, was hiding a complex of vertical cracks passing through the stones and continuing along the mortar joints (fig. 6). This plaster did not probably originate with the church, but only attached in the sixties as for other churches in Noto. Furthermore, after the removal, it was possible to detect that it was not adhering well to the masonry, also an adequate mechanical preparation of the support was absent; in fact, a dirty patina on the stones under the plaster itself was found. The plaster was particularly thin

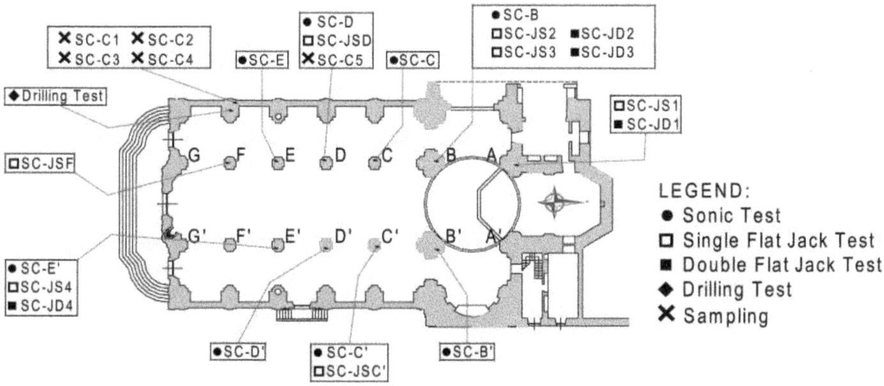

Figure 5: Crocifisso Church plan and localisation of the double flat-jack tests.

(fig. 6d) and completely detached in the area in which deep cracks were distributed.

The pillars, as in the case of the Noto Cathedral, were built with an external leaf made of "Noto stone" regular blocks and internal filling with rubble masonry; the technique of construction is similar to the one of the Cathedral pillars, but the internal rubble seemed to be better built. After the recognition of the damages, the removal of the plaster from all the pillars up to a height of 3 m was planned in order to survey the crack pattern.

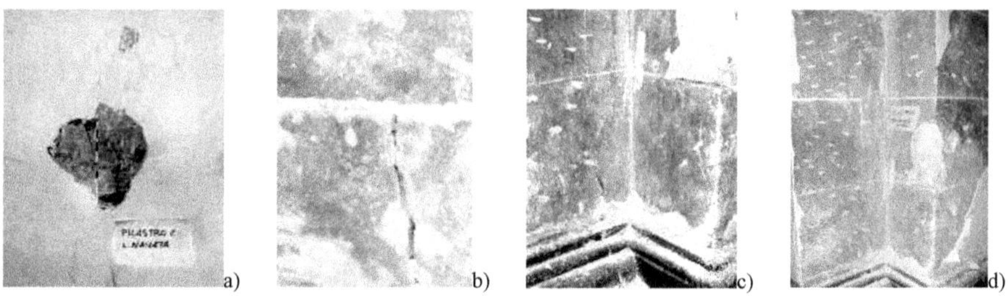

Figure 6: (a) The plaster showed only very fine cracks; (b, c) damages visible after the plaster removal; d) a very thin plaster (4–5 mm thick) made with lime-gypsum mortar covers the pillars, hiding the cracks.

Figs. 7–9 show examples of the crack pattern of the pillars. As it is possible to observe in fig. 8, the cracks are diffused and intersecting the whole view, with a concentration in the corners. This is a typical effect of damage under high compression. In some cases, the cracks were filled by mortar and the corners reconstructed (figs. 7 and 8) with mortar. This fact suggests that the crack pattern was present before the plaster was applied. Fig. 9 shows other details of the damages.

The strengthening design was urgently changed and the pillars were urgently confined by stainless steel, cross-shaped reinforcements confining the pillars at every two courses.

3.1.2 NDT and MDT on the pillars

The coring and the boroscopy observation, which together with the stone disposal survey gave information on the section morphology, revealed a multiple leaf masonry (see also fig. 5). The internal masonry, as mentioned above, is a rubble one, composed by a rather weak mortar, pieces of limestone (fig. 10a, fig. 11).

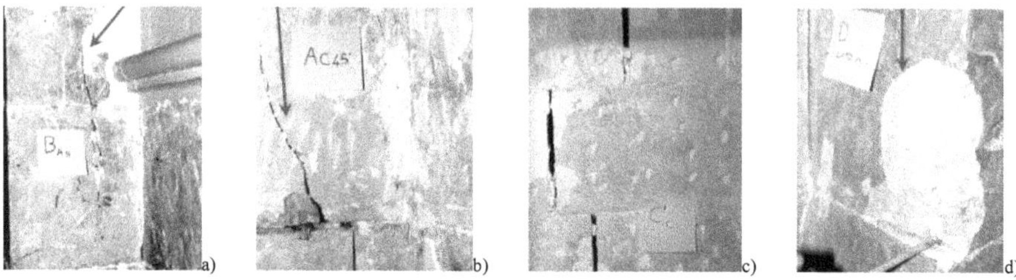

Figure 7: Example of damages: (a, b) vertical cracks on the stone-blocks, (c) empty vertical joints (d) lack of material.

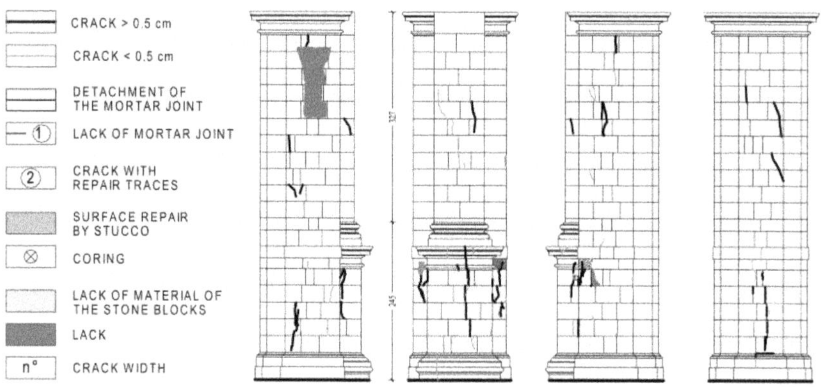

Figure 8: Crocifisso Church: crack pattern of the base of pillar C1.

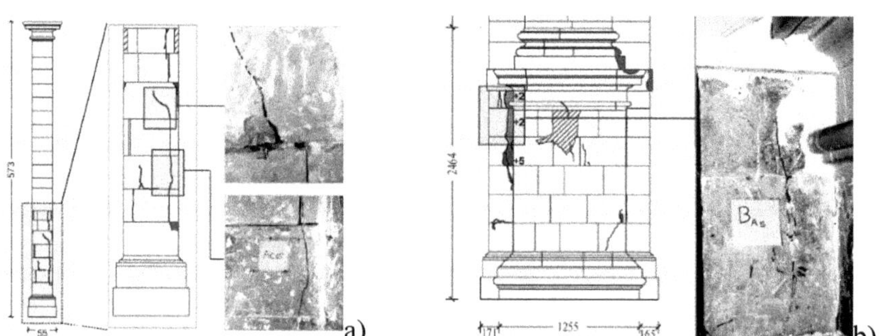

Figure 9: Crack pattern survey and detail of pillars B (a) and C (b) (fig. 5).

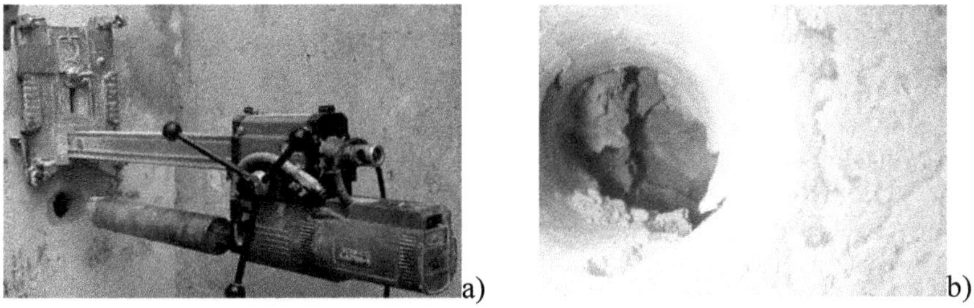

Figure 10: a) Core sampler applied to the masonry structure and b) view of the internal structure made by rubble material and rather weak mortar.

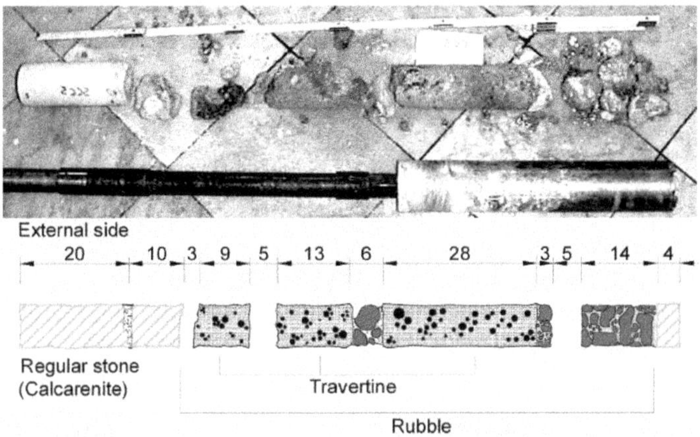

Figure 11: Drilled core of pillar C (fig. 5) and re-construction of the sample.

3.1.2.1 Pulse sonic tests Pulse sonic tests were carried out in order to detect the pillars characteristics. Fig. 12 shows typical sonic test results by transparency. The sonic velocities are higher in co-incidence with the external stone leaves and lower when the impulse crosses the internal rubble filling, as expected. These results confirm what was found by coring (fig. 11).

The sonic velocity also is lowered by the presence of cracks, when they are perpendicularly intersected. In fact, the lowest average velocities (SC-D1) concern the pillar called D (fig. 5), characterised by the worst crack pattern situation. It was also possible to notice that, in the case of highly cracked external leaf, the velocity became lower.

The sonic tests once again confirmed their efficiency in detecting difference in density of the material (i.e. rubble filling of pillars was confirmed) (fig.13). This possibility avoids the use of destructive and invasive tests also allowing for the investigation on large surfaces of masonry.

3.1.2.2 Flat jack tests Flat jack tests, single and double, provided interesting information. First of all, the state of stress was lower and non symmetrical in the pillars supporting the dome, with a value of 0.76 and 1.04-1.09 MPa,

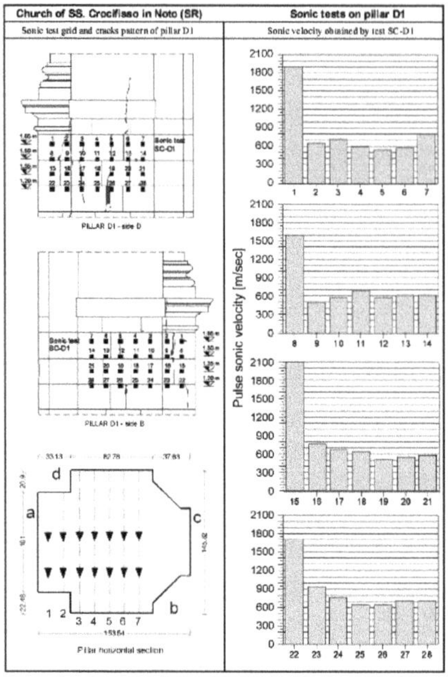

Figure 12: Crocifisso Church: results of sonic tests applied to pillar D1.

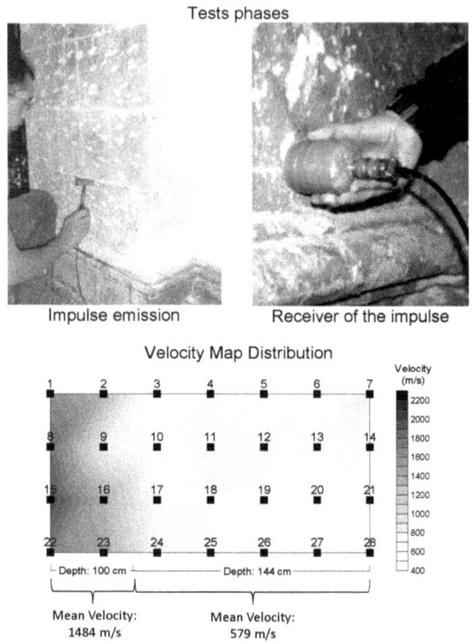

Figure 13: Sonic tests carried out on pillar D1 represented by velocity map.

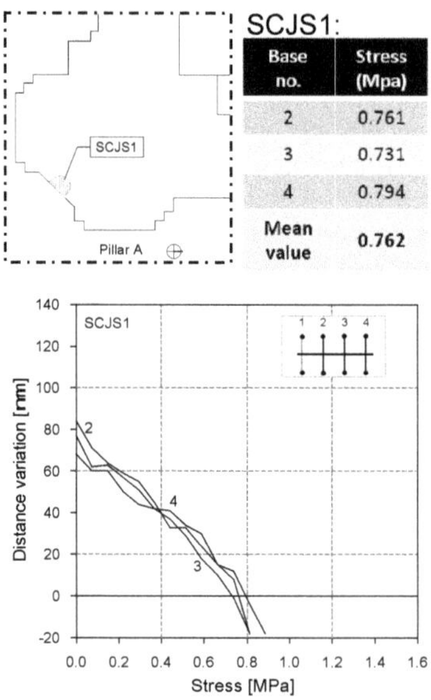

Base no.	Stress (Mpa)
2	0.761
3	0.731
4	0.794
Mean value	0.762

Figure 14: Single flat jack test SCJS1 carried out in apse pillar A sustaining the dome.

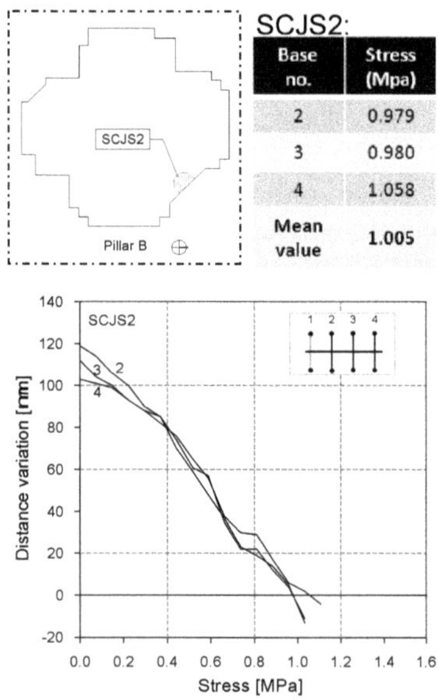

Base no.	Stress (Mpa)
2	0.979
3	0.980
4	1.058
Mean value	1.005

Figure 15: Single flat jack test SCJS2 carried out in apse pillar B sustaining the dome.

respectively (fig. 5) in the apse pillar (SC-S1) and in the pillar toward the transept (SCJS2) (figs. 14 and 15). The lower stress in the apse pillar is justified by the presence of wall portions, which support part of the loads. The rather lower stresses measured on these pillars are also due to their large dimensions compared to the other main nave pillars (fig. 5).

Higher stress values, 1.39 MPa, were measured (SC-JS4) on a pillar called E1 (fig. 5), while the other pillars have a lower stress (0.83–0.99 MPa). The stress-strain diagrams show consistent deformability of the masonry with rather low values of the elastic modulus and hence also of the strength. Furthermore, the high diffusion of cracks caused the interruption of most of the tests. The double flat jack-tests carried out within two stones (LVDT 5,6,7) and across a mortar joint (LVDT 1,2,3,4) (fig. 16), show a totally different deformation under the same stress when it is measured on the stones or across the joint.

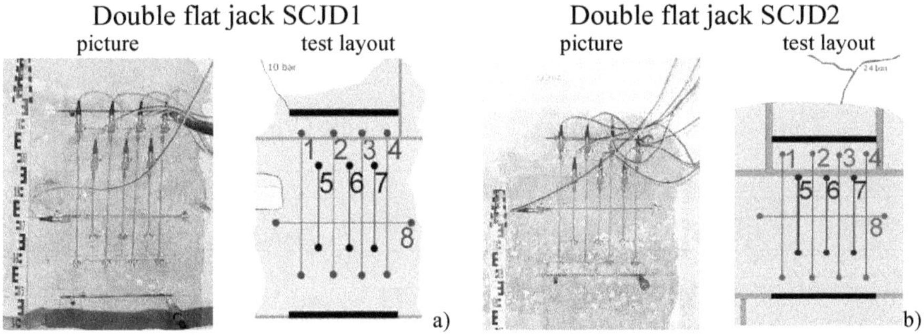

Figure 16: Geometry of the double flat jack-tests SCD1 (a) and SCD2 (b). The acquisition are carried out within a stone (LVDT 5, 6, 7) and across a mortar joint (LVDT 1, 2, 3, 4).

In figs. 17 and 18 (SC-D1, SC-D2) these two different types of behaviour are shown in the stress-strain plot as average of LVDTs 1, 2, 3, 4 (fig. 16). It is thus confirmed that there is a greater deformation in the joint when a certain state of stress is reached. Furthermore, the plot of fig.17 shows a weaker behaviour of the masonry compared to SC-D2 (fig. 18).

Nevertheless, in both cases the state of stress measured by the single flat-jack is outside the elastic range of the material. This explains also the detected damages.

3.1.2.3 Material characterisation The used limestone, the "Noto Stone", is characterised by low mechanical properties especially in the case of saturated samples, as was revealed by the compressive and splitting tests. Table 1 shows the results of the mechanical results carried out on stones sampled from the walls of the Church. The mortar sampled from the pillars was submitted to chemical and optical tests and was defined as a hydrated lime based mortar. Table 2 shows the results of chemical analyses carried out on mortar samples.

WIT Transactions on Engineering Sciences, Vol 80, © 2012 WIT Press
www.witpress.com, ISSN 1743-3533 (on-line)

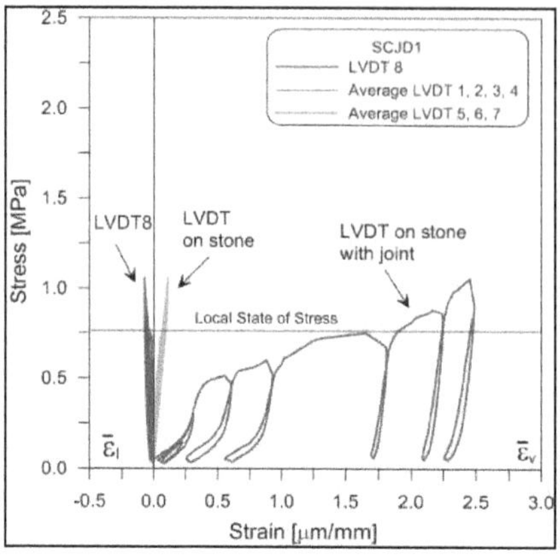

Figure 17: Flat-jack test result (test SC-D1 in fig. 5).

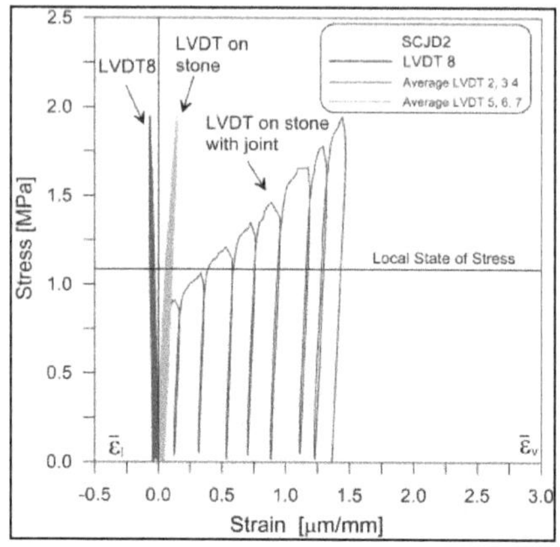

Figure 18: Flat-jack test result (test SC- D2 in fig. 5).

Table 1: Mechanic characteristic of the Crocifisso stones.

	Compressive Test			Brazilian Test	
Sample	Ultimate Stress [MPa]	E [MPa]	ν (30-60%)	Sample	Ultimate Stress [MPa]
Dry condition					
SCC2i	16.1	11000	0.15	SCC5	2.27
SCC4i	19.1	12150	0.17	SCC3	2.09
Saturated					
SCC2e	10.7	7370	0.19	SCC3	1.27
SCC3i	12.5	10000	0.16	SCC4e	1.09
SCC5e	10.3	9650	0.16		

Table 2: Results of chemical analysis on two mortar samples (courtesy of G. Baronio).

	Internal mortar sample	**Joint mortar sample**
Determinations	**%**	**%**
Silica (SiO_2)	3.40	2.57
Alumina (Al_2O_3)	0.57	1.30
Ferric Oxide (Fe_2O_3)	0.50	0.90
Calcium Oxide (CaO)	50.27	51.12
Magnesium Oxide (MgO)	0.30	0.30
Sodium Oxide (Na_2O)	0.70	0.47
Potassium Oxide (K_2O)	1.79	0.93
Sulphuric Anhydride (So_3)	0.15	0.20
Loss of Ignition	42.03	41.90
Carbon Dioxide (CO_2)	40.04	40.98
Insoluble residue	3.36	2.70
Soluble Silica	1.04	1.03
Chlorites	0.34	0.36
Bulk Density	1368 Kg/m^3	-

3.2 Tests on timber structures

The tests carried out on the timber elements of the roof were aimed at assessing the material density related to the humidity and to the timber species. Initially, the thermo-hygrometric conditions at the testing points were recorded (fig. 19a). Subsequently, the ultrasonic characterization was carried out in cross-section mode (fig. 19b).

Finally, penetrometric tests were performed using an apparatus, commercially designated Resistograph (RESIF400®) (fig. 19c), that measures the resistance opposed by the timber to the penetration of a pin which is rotating and penetrating at constant speed. This parameter is related to the material density, which is, of course, influenced by the presence of knots, irregularities, decay, etc (fig. 20).

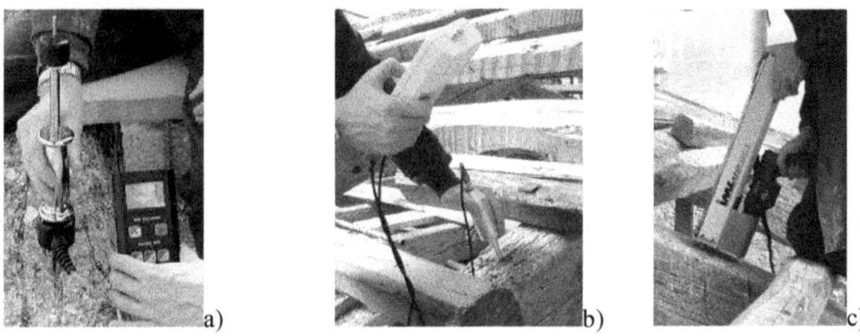

Figure 19: Testing the timber roof: thermo-hygrometric survey (a), ultrasonic test (b), penetrometric test on a truss element (c).

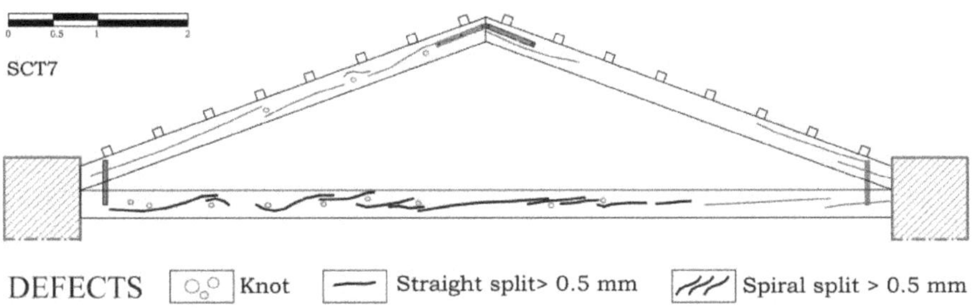

Figure 20: Example of survey of a typical defect of a timber structure.

For each truss structure, three profiles were collected, one on the horizontal element, one on each inclined element and another on the secondary elements as shown in fig. 21(a). As an example, a profile obtained at the end of a horizontal element is shown in fig. 21(b). In general, very few profiles, localized at the ends of the horizontal elements, showed internal discontinuities.

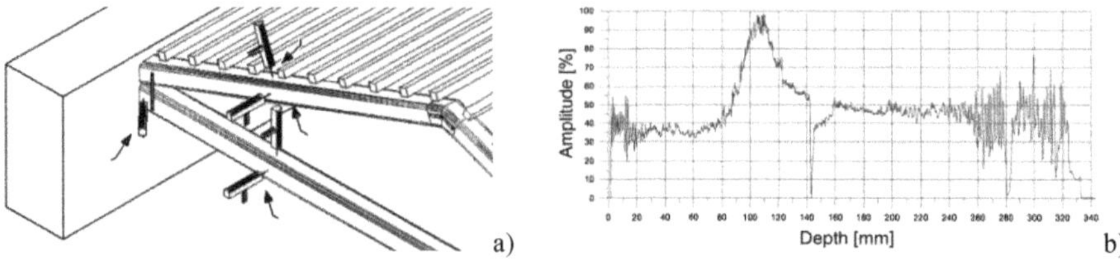

Figure 21: Penetrometric tests: localization of the tests (a); amplitude vs. depth profile (b).

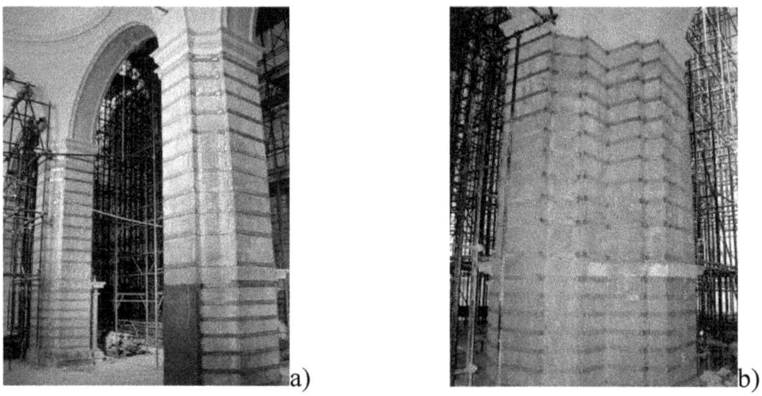

Figure 22: View of the steel confinement applied to a) the pillars of the nave and b) the pillars of the dome.

4 Conclusions

The investigation carried out on the structures of the Crocifisso church was particularly important for both the timber and the masonry structures. For the former, it was decided that they could be preserved with few repairs. For the pillars, the situation was considered so difficult that it was suggested to carry out a repair intervention. The case in fact was very similar to the one of the Noto Cathedral. A partial collapse could take place under a next earthquake or due to the long term behaviour of masonry without seismic events given the diffused damages shown in the form of vertical diffused cracks of the studied pillars.

Even if in similar cases, it is suggested to monitor the structure for some years, in this particular case the work was done soon. The intervention was carried out on all the pillars by confining them at different heights with appropriate steel confinements.

The results of visual inspection, survey and NDE tests once again showed the importance of a well-designed investigation to detect and understand the damage and its causes.

Acknowledgements

The authors wish to thank the parish priest who supported the experimental investigation. The research was financed by the Genio Civile of Syracuse. C. Arcadi, M. Antico, M. Cucchi, M. Iscandri D. Penazzi, C. Tedeschi, C. Tiraboschi are gratefully acknowledged for their technical support; many thanks are also due to the students. Eng. Messina is gratefully acknowledged for his support from the Genio Civile.

References

[1] De Benedictis, R., Tringali, S., Gavarini, C., Binda, L. & Baronio, G., Methodology applied to the removal of the ruins and to the survey of the remains after the collapse of the Noto Cathedral in Sicily. *Proc. of the 6th Int. Conf. Structural Studies, Repairs and Maintenance of Historical Buildings, STREMAH 99*, Dresden, Germany, pp. 529-538, 1999.

[2] Binda, L., Saisi, A. C., De Benedictis, R. & Tringali, S., Experimental study on the damaged pillars of the Noto Cathedral. *8th Int. Conf. on Structural Studies, Repairs and Maintenance of Heritage Architecture STREMAH 03,* Halkidiki, Greece, pp. 89-98, 2003.

[3] Binda, L., Saisi, A. & Tiraboschi, C. Application of sonic tests to the diagnosis of damaged and repaired structures. *Non Destructive Testing and Evaluation International,* **34**(**2**), pp. 123-138, 2001.

[4] Binda, L., Gatti, G., Mangano, G., Poggi, C. & Sacchi Landriani, G., The collapse of the Civic Tower of Pavia: a survey of the materials and structure. *Masonry International*, **6**(**1**), pp. 11-20, 1992.

[5] Binda, L. & Anzani, A. The time-dependent behaviour of masonry prisms: an interpretation. *The Masonry Society Journal*, **11**(**2**), pp. 17-34, 1993.

[6] Binda, L., Saisi, A., Messina, S. & Tringali, S., Mechanical damage due to long term behaviour of multiple leaf pillars in Sicilian Churches. *Proc. of the III Int. Seminar: Historical Constructions*, pp. 707-718, 2001.

[7] Binda, L., Saisi, A. & Tiraboschi, C. Investigation procedures for the diagnosis of historic masonries. *Construction and Building Materials,* **14**(**4**), pp. 199-233, 2000.

The textile block system: seismic analysis and upgrading

A. P. Vargas & G. G. Schierle
USC School of Architecture, Los Angeles, CA, USA

Abstract

The textile block system is a unique structural system created by Frank Lloyd Wright in the early 1920s, before any seismic regulations existed in California. The first four houses, all located in California, have been deteriorating severely due to seismic and environmental effects. Previous research and interventions to preserve them have not been completely successful. It is therefore important to carry out an analysis of the seismic vulnerability of these structures, and explore ways of strengthening them to comply with current seismic requirements. This paper demonstrates that the textile block system can be upgraded to current structural standards for seismic safety based on the International Building Code, IBC 2003 and codes for historical structures, including the California Historical Building Code and the Federal Emergency Management Agency (FEMA). The Freeman House is used as a case study. The allowable stress design method (ASD) is used for the analysis. The paper also discusses procedures and construction methods to strengthen existing textile block structure. The methodology presented may be used for seismic upgrading of other historic structures as well.
Keywords: textile block, Frank Lloyd Wright, ASD, allowable stress, shear walls, base shear, seismic analysis, seismic upgrading.

1 Introduction

With the recognition of Frank Lloyd Wright as a great architect and designation of many of his buildings as historical monuments, there has been increasing emphasis placed on the preservation and conservation of his masterpieces.

This paper aims to analyze and upgrade with an affordable and non destructive method the Freeman House, designed by Frank Lloyd Wright in

1924, as a case study for testing the textile block structural system, and compare that system with those used in three other textile block system houses: the Millard, Storer and Ennis-Brown Houses.

Figure 1: Millard House/block detail, Pasadena, CA. Freeman House/block detail, Hollywood, CA. Photos Angela Paola Vargas, 2004.

Figure 2: Storer House/block detail, Hollywood, CA. Ennis-Brown House/block detail, Hollywood, CA. Photos Angela Paola Vargas, 2004.

1.1 The textile block – an overview

In order to be able to assess damage to the textile block houses and propose solutions for preservation, it is important to first thoroughly understand the original components and structure of the textile block system. After the Millard House, Frank Lloyd Wright set about developing a way of strengthening the system. An important feature of his original system was that the blocks are stacked on top of each other without a mortar setting bed. Therefore, he improved the system by adding a network of inter-block joints, filled with grout and steel reinforced rods, similar to adobe construction in strength and resistance.

The original block was a 16"×16" concrete tile, with a 1½" diameter semi-circular channel running along each of the four sides, such that when two blocks are joined side-by-side, a circular channel is created, through which a steel bar is run. For the Freeman House each block is of the same pattern (fig. 3). The blocks were created by pouring concrete mixture into molds, enabling the repetition of form, and reflecting the mechanization trend of that period in America. The textile blocks are reinforced by manufactured steel rods, which were run through the circular channels between the concrete blocks. Mortar was then poured into the channels, to bond the steel bars with the blocks. The steel grid in essence functions as a lateral system to resist shear, rather than spanning the whole structure.

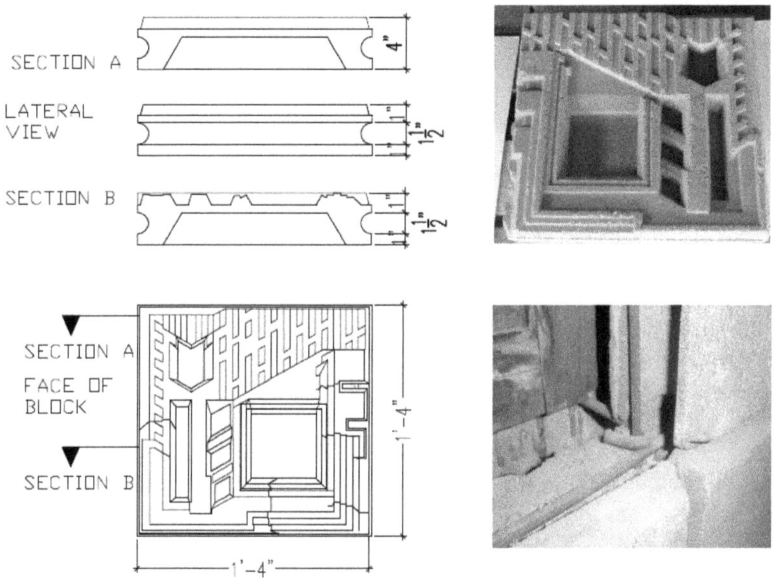

Figure 3: Freeman House, block dimensions, CAD – drawing/photos – Angela Paola Vargas, 2004, interpretation from archives.

Figure 4: Ennis-Brown residence, photos – Angela Paola Vargas – 2004.

The types of damage typically found can be grouped into the following categories, according to this report: deterioration or crumbling, spalling and ring fractures, erosion or weathering, and seismic shear cracks.

Figure 5: Typical damage, photos – Angela Paola Vargas – 2004.

1.2 The textile block system and building codes

The Freeman House was built in 1924 and it is currently listed in the National Register of Historic Places. Therefore, any changes or restorations must comply with current building codes such as the 1997 Uniform Building Code (in order to comply with the State of California and the City of Los Angeles), the Secretary of Interior's Standards for Rehabilitation, and the State Historic Building Code.

This analysis uses the fundamental assumptions, formulae and design procedures for the Allowable Stress Design (ASD) method. This method was selected for the analysis of the Freeman House following the recommendations provided by the California Historical Building Code, title 24, part 8 for existing structures. It is a non-invasive method, which is ideal for the Freeman House, a historical monument and an eligible National Landmark. In addition to the ASD method, the IBC 2003 and the USGS outline specific guidelines and coefficients that are used for this analysis.

2 Seismic analysis of the textile block system

2.1 Seismic examination of a selected area of the original Freeman House – allowable stress design

There are three basic assumptions in the analysis of the original Freeman House. Firstly, it is assumed that the walls act as ordinary reinforced masonry, according to the IBC 2003. This masonry is weak in shear walls. Secondly it is necessity to define the character of a structure as belonging to a category established by the IBC, through definition of materials, ultimate strength or yield strength, and a factor of safety. Thirdly, the structure is composed of steel-reinforced blocks, but further reinforcing steel is necessary to provide the shear strength and ductility necessary to resist seismic forces.

2.1.1 Block analysis

In order to measure the overall dead load of the selected area of the Freeman House, the basis of the Allowable Stress Design method, it is necessary to calculate individual block weight, which has been found to be on average 85 pcf (pounds per cubic foot).

Table 1: Block weight calculation.

Block Weight (volume)	85 pcf
Block Weight (area) psf	$1.03\text{ft}^3 \times 85$ pcf / 1.32 = 51.8 psf
Block Weight (area) psf	52 psf

2.1.2 Dead load calculation

The ASD method is based on dead load, as defined by codes. The total dead load of the original section of the Freeman House is required to compute the base shear, which in turn is used to define the distribution of forces per level.

Table 2: Dead load (DL).

Roof DL	W = 97 k
Ground Floor Level 1 DL	W = 118 k
Total Dead Load	$\sum$W = 215 k

2.1.3 Determining the Freeman House seismic factor using IBC tables and the USGS earthquake hazard parameters

In order to determine the seismic factor, the Design Spectral Acceleration (S_{DS}) for the building needs to be defined. The first step is to identify the site class, using the IBC tables for the Freeman House, the default category D is used. The USGS Earthquake Hazard Parameters, define for each site probabilistic spectral accelerations, defined as S_S (for low-rise) and S_1 (for high-rise) structures.

USGS spectral accelerations for seismic design are based on 2% probability of exceeding (PE) in 50 years. Based on the spectral acceleration S_S, IBC defines Design Spectral Acceleration S_{DS} and the seismic coefficient C_S as

$$C_S = I\, S_{DS}/R$$

where

I = Importance factor
R = Reduction factor

The seismic coefficient C_S used to compute base shear:

$$V = C_S W$$

where W = dead load

The S_{DS} graphs (fig. 6) [8] provide a streamlined method to define S_{DS}.

The seismic parameters of the Freeman House are:

S_S = 205%, as decimal S_S = 2
For site class D, S_{DS} = 0.95
Importance factor I = 1
Masonry response factor R =2
Base shear coefficient $C_S = I\, S_{DS} / R$ = 1(0.95)/2 = 0.45.

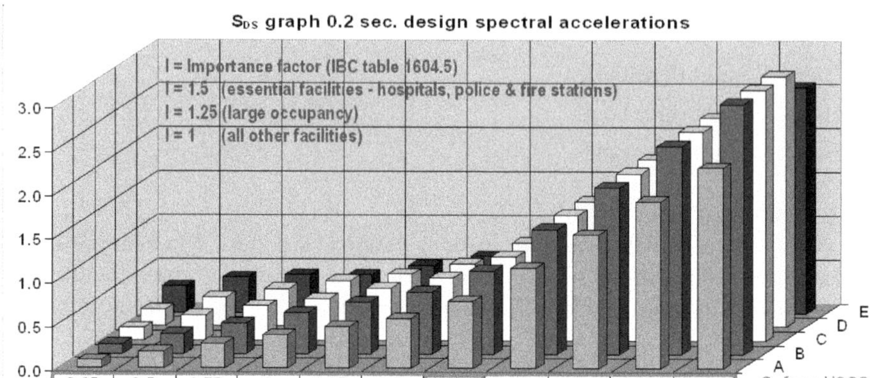

	S_S from USGS: 0.25	0.5	0.75	1	1.25	1.5	2	3	4	5	6	S_{DS}
A	0.10	0.19	0.29	0.38	0.48	0.57	0.76	1.14	1.52	1.90	2.28	hard rock
B	0.12	0.24	0.36	0.48	0.60	0.71	0.95	1.43	1.90	2.38	2.86	rock
C	0.14	0.29	0.39	0.48	0.60	0.71	0.95	1.43	1.90	2.38	2.86	dense soil
D	0.19	0.33	0.43	0.52	0.60	0.71	0.95	1.43	1.90	2.38	2.86	stiff soil
E	0.30	0.40	0.43	0.43	0.54	0.64	0.86	1.29	1.71	2.14	2.57	soft soil

Figure 6: S_{DS} graphs. (Site classes A-E, default site class D used.)

2.1.4 Base shear V

Base shear, the total lateral force at the building base is computed as follows:
Dead load W = 215 k + 20 k (estimate for seismic upgrades), W = 235 k
Adjustments per IBC: Ordinary masonry factor = 1.5
Existing building factor = 0.85

Base shear V

$$V = C_S\, W\, (1.5)\, (0.85) = 0.48\, (235\ k)\, (1.5)\, (0.85) = 144\ k$$

2.1.5 Force and shear distribution per level

Force distribution per level is computed as

$$F_X = C_{vx} V$$

$$C_{VX} = w_x h_x^k / \sum_{i=1}^{n} w_i h_i^k$$

Shear distribution per level is computed as

$$V_X = \sum_{i=x}^{n} F_i$$

Force and shear distribution per level are tabulated in table 3. Shear per floor area is tabulated in table 4.

Table 3: Force and shear distribution.

Force and Shear Distribution per Level						
Level	Wx	hx	Wxhx	Wxhx ∑ Wihi	Fx = V (Wxhx / ∑ Wihi)	Vx= ∑Fx
Main (Level 1)	97k	18'	1,746k'	0.6	144k (0.6)=86k	86k
Lower Level 0)	118k	10'	1,180k'	0.4	144k (0.4)=58k	144k
			∑Wihi= 2,926k'			

Table 4: Shear per floor area (square foot).

Shear per Floor Area per Level				
v = V/A	V Per Level	A	V/A	v (psf)
Level 1	86k = 86,000 #	1204 sf	86,000/1,204	71 psf
Level 0	144k=144,000 #	1208 sf	144,000/1,208	119 psf

2.1.6 Rebar size and spacing

Rebar size and spacing are computed as follows:

The walls are 8" thick, with one inch of hollow space. Therefore, for calculation purposes, they are assumed to be 7" thick. The original rebars are ineffective as reinforcement for the structure. The seismic upgrading assumes:

- Grade 60 rebars with allowable stress of Fs =24000 psi.
- Masonry specified compressive strength of fm =1500 psi
- Allowable masonry shear stress of Fv =24 psi.
- Number 4 rebars (0.5" diameter) cross section area Av = 0.2 in^2.

Rebar spacing S is defined by the following formula:

$$S = Av\ Fs / (Fv\ b)$$

where Av is the rebar cross-section area, Fs is allowable rebar stress, Fv is allowable wall shear stress, and b is wall thickness. The space required between bars is computed as:

$$S = (0.2)\ (24000) / (24)\ (7") = 28"$$

Considering the 8" masonry modules the bar spacing used is S = 16".

2.1.7 Determining the required effective wall length per level

The following formula determines required length per level (d'), which is based on the base shear per level, divided per the allowable shear stress Fv by the wall thickness b.

Table 5: Required shear wall lengths.

Level 1	Converting to feet	
d' = V / (Fv b)	512" / 12 = 43"	
d'=86,000#/24) (7")	L=d+8"	
d'=86,000# / 168	L= 43.66'	Use L=44'
d'=512"	Use 33 modules of 16"= 44' each way	
Level 0	Converting to feet	
d' = V / (Fv b)	857" / 12 = 71'	
d'=144,000#/24(7")	L=d+8"	L=71'+0.66
d'=144,000# / 168	L=71'+0.66"	L=72'
d'=857"	Use 54 modulus of 16" = 72' each way	

References

[1] Abrams, D. P., Seismic response patterns for URM buildings. *The Masonry Society Journal*, **18(1)**, pp. 71-78, 2000.

[2] Applied Technology Council, *ATC-20: Procedures for Post-earthquake Safety Evaluations of Buildings*. R. P. Gallagher Associates, Inc., San Francisco, 1989.

[3] Applied Technology Council, *ATC-21-1: Rapid Visual Screening of Buildings For Potential Seismic Hazards: Supporting Documentation.* Redwood City, California, 1988.

[4] Blake, P., *The Master Builders*. Toronto, Canada and the United States of America. First published in the Norton Library, 1976.

[5] Chusid, J. M., *Historic Structure Report: Samuel and Harriet Freeman House, Hollywood, California, 1924,* Prepared for the School of Architecture, University of Southern California, Los Angeles, California, 1989.

[6] FEMA 356, Chapter 7: Masonry, Chapter 11: Architectural, Mechanical and Electrical Components, and Appendix A: Use of this Standard for Local or Directed Risk Mitigation Programs.

[7] Schierle, G. G., *Structures in Architecture*, School of Architecture, University of Southern California, 2006.

[8] Moor, A., *California Textile Block*, PRC: London, New York, 2002.

[9] Murray, M., A. PE, C.E.O, Specialist in Structural, Repair Restruction Corporation, Telephone interview in Sedalia, Colorado, November 18, 2004.

[10] Smith, K., *Frank Lloyd Wright: Hollyhock House and Olive Hill: Buildings and Projects for Aline Barnsdall*, Rizzoli: New York, 1992.
[11] Strengthening of Masonry with FRP (Fiber Reinforced Polymer) Bars, ICCI, San Francisco, 2002 http://campus.umr.edu/rb2c/publications /conf/2002/strngthng_masonry.pdf
[12] Sweeney, R. L., *Wright in Hollywood: Visions of a New Architecture*, The Architectural History Foundation and the Massachusetts Institute of Technology: New York, 1994.

Damage assessment and retrofitting of Marchesale Castle (San Giuliano di Puglia)

R. Cami[1], C. Alessandri[2], M. Indirli[3] & A. Tralli[1]
[1]*Department of Engineering, University of Ferrara, Italy*
[2]*Department of Architecture, University of Ferrara, Italy*
[3]*ENEA, Bologna, Italy*

Abstract

After the earthquake that struck the Molise Region (Italy), some ENEA researchers were involved in several post-seismic activities in San Giuliano di Puglia. Ancient and notable masonry cultural heritage structures (MCUHESs) suffered severe damage or partial collapse, including the old Marchesale Castle. This paper deals with a repairing proposal for the building, developed in the framework of a PhD thesis; diagnostics and dynamic investigations results are elaborated, numerical calculations are provided and interventions are suggested in agreement with the existing code. The main purpose is to enhance the building's seismic performance, avoiding conflicts between conservation and safety.
Keywords: seismic behavior, vibrations, assessment, retrofitting.

1 Introduction

1.1 The seismic event of October 31st, 2002

A moderate earthquake struck Molise (Italy) on October 31st, 2002. The first shock (M 5.4) was followed by another (M 5.3) on the following day. Spread damage was evident in San Giuliano di Puglia, a small town located 5 km from the epicenter. The maximum intensity at the site was estimated to be VIII-IX MCS, observed during both the 1456 and the 2002 events. The damage was not uniformly distributed inside the San Giuliano narrow area, but characterized by different levels of seismic hazard and structural vulnerability. ENEA experts took part in these activities following the seismic event: a) the emergency, under the coordination of the Civil Defense; b) the post-emergency, carrying out a detailed evaluation of damage, drafting the demolition plan, ensuring safe conditions in the buildings to be repaired, and working towards allowing residents to safely reenter their non-damaged houses; c) the reconstruction plan of San Giuliano; d) the technical-scientific advice to the Office of the Public Works Ministry for reconstruction and rehabilitation projects, including Marchesale Castle [10–12].

1.2 Seismic input

Before the earthquake, San Giuliano was not classified as a seismic zone, but subsequently it has been included in zone 2 (maximum Peak Ground Acceleration, PGA, equal to 0.25g) in the seismic reclassification of Italy [16]. The Civil Defense appointed a technical commission to provide the seismic microzoning of San Giuliano. The study reflects zones with different geology, topography and seismic local amplification. The 1456 earthquake (M 6.6, MCS XI) hit a large area of South-Central Italy; the event probably consisted of a long series of shocks, due to the simultaneous activation of several seismic sources, overlapping at least three areas encircling different epicenters [10–12].

1.3 Structural vulnerability of the San Giuliano historical center

The medieval center (fig. 1), interesting from an historical and architectural point of view, was investigated in depth. Despite the low local seismic amplification, it presented medium-severe damage and partial collapses. The Marchesale Castle (fig. 2) suffered heavy damage as well. All the area was generally characterized by high vulnerability and the most usual collapse mechanisms were wall failure with typical cross cracks but also out-of-plane overturning [10–12].

Figure 1: San Giuliano historical center.

Figure 2: Damage to the castle.

1.4 Design codes and requirements

In many scientific studies, it is well-established that antiseismic interventions on MCUHESs located in seismic areas must harmonize protection and conservation [8, 9], learning from periodical earthquake lessons [6, 7]. Furthermore, the new Italian seismic code [16] and the related guidelines for cultural heritage [17] contain specific requirements to follow. In the authors' opinion, the approach to MCUHESs structural improvement can be summarized in the following statements: a) since repair is much more difficult to be carried out than that related to modern r.c. or steel structures, interventions can deviate from the antiseismic design criteria adopted for ordinary buildings; b) in relation to the state limit analysis, interventions must be defined as a "controlled structural improvement", i.e. acceptance of an antiseismic protection level lower than required, reducing invasivity; at the same time, great attention must be given to the building final purpose, in order to manage with accuracy the "Importance Factor" used in seismic standards (higher for public and strategic constructions [13]); c) the improvement effectiveness must be quantified, evaluating PGA levels generating local collapse mechanisms; d) detailed surveys and investigation campaigns are mandatory for MCUHESs, whose characteristics are frequently not well known; e) it is necessary to undertake the rehabilitation design in a specific way, since the use of standardized procedures is not possible; f) the observance of the "regola dell'arte", i.e. unwritten construction rules for masonry elaborated by architects and bricklayers during centuries of work practice, is fundamental for protection, conservation and restoration; g) the use of modern techniques and materials can be very useful to reduce seismic vulnerability, but it must be historically correct, compatible and mechanically effective.

2 Description of Marchesale Castle

Historical information on Marchesale Castle (typical example of Southern Italian architecture) is very scarce. Around the first millennium A.D. a fortress with three towered doors was already built. The 1456 earthquake destroyed most of the medieval center, then reconstructed. Until present times, inhomogeneous works formed layers, so that reworked parts are distinguished from almost untouched ones. The castle consists of several interconnected blocks; taking as reference the internal court ground, 4 upper and 3 lower levels envelope the hill slope (fig. 3). Multi-leaf stone-masonry walls are characterized by poor materials, irregular morphology and voids presence (concentrated in a loose internal core), with undoubted structural problems (internal layer weakness, mortar deterioration in the external joints, lacking of connections among external leaves). Therefore, the building is sensitive to "brittle" collapse mechanisms, both under vertical and horizontal loads, by layers detachment and out-of-plane ejections. The tower (fig. 4) represents the most precious element of the entire complex. It is a squared-plan building, with buttresses on the West side and a Southern entrance portal (with a barrel vault) to the internal court. Most of the castle horizontal structures are generally wooden or steel-tied floors. Original stone vaults, with different geometries, can be found. The tower thick basement (2.90 m) shows pronounced tapering with a height up to 1.20 m, and irregular masonry fabric in the façades, made by heterogeneous materials (mainly corner squared stones, variously sized river pebbles and brick insertions), as the majority of the castle walls. In the tower front it is still possible to identify pretty genuine architectural evidences, the so called "buche pontaie" (small holes used to provide supplemental defense wooden scaffolding) and the twice-line triangular pigeon houses located under the roof. Other aspects are: generally deteriorated mortar courses, plaster traces, brick fillings of old openings. After the seismic event, the tower top was tied up with steel cables and covered with a steel sheet.

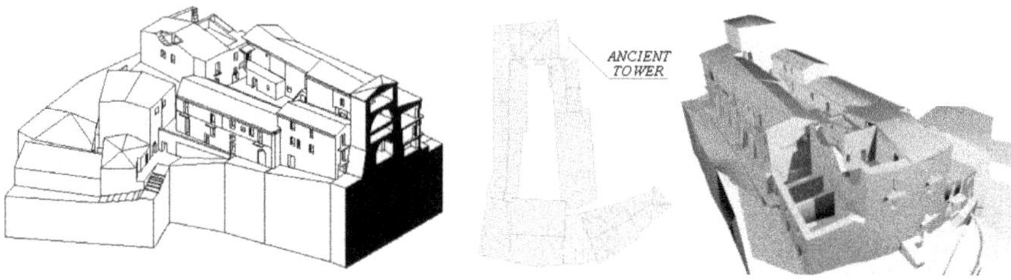

Figure 3: Geometry of Marchesale Castle.

Figure 4: The old Marchesale Castle Tower.

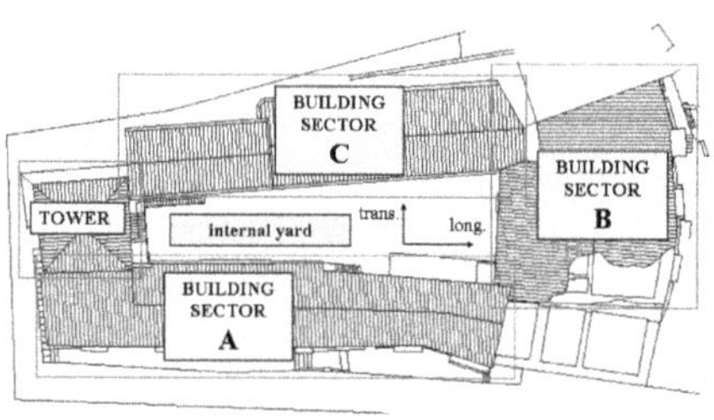

Figure 5: Building sectors characterized during the dynamic tests.

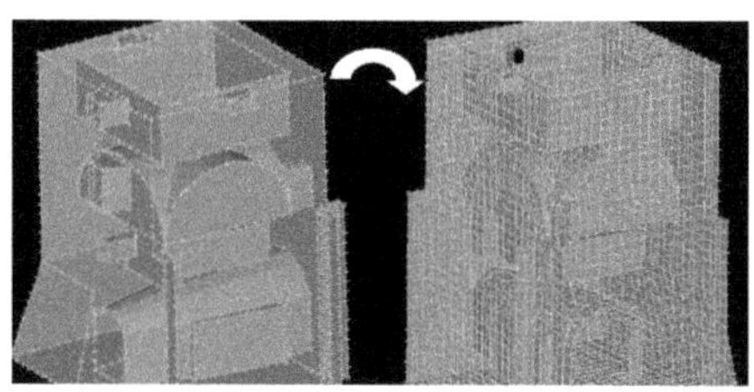

Figure 6: FEM implementation from geometric survey data.

3 Material diagnostics

A detailed description of the work carried out on the Marchesale Castle is given in [12]. Flat jack tests measured local compression (1.0–4.0 kg/cm^2) and Elasticity modulus (2000–3500 MPa); the compressive resistance can be estimated around a value of 15.0–20.0 kg/cm^2. Shear pull-out tests (generally done in the same positions as the flat jack tests in order to correlate masonry sliding properties with local axial compression) gave values around 2.0–3.5 kg/cm^2. Borehole coring with video endoscopy tests (performed on elevation and foundation walls with particular attention to the tower) showed, in most cases, a poor internal core (small stones, lack of mortar and voids presence). Sonic pulse velocity tests detected the presence of voids, cracks and damage patterns.

4 Dynamic tests

Ambient vibration tests [3] have been carried out in order to characterize various castle sectors (fig. 5). For sector A, a modal shape (5.8 Hz) has been identified in the transversal direction; another peak (4.5 Hz) is probably due to the presence of the adjacent tower. In the longitudinal direction, a frequency of 10.7 Hz has been detected, but also less evident peaks (8.0 and 8.8 Hz). For sector B, a frequency of 5.8 Hz has been noted in the transversal direction (maybe a torsion mode); another value (4.5 Hz) seemed of translational nature. In the longitudinal direction, frequencies of about 6.0 and 8.5 Hz have been found, but also other peaks between 10 and 15 Hz. The tower behaviour is clear: two main frequencies (4.5 Hz transversal and 5.1 Hz longitudinal) seemed both bending modes; a third value (7.5 Hz) is probably associated to torsion.

5 Numerical models

The use of finite element models (FEMs) for MCUHESs, implemented by data coming from tests, can be helpful to understand the static and dynamic behavior, identify modal frequencies, evaluate stress, strain, displacement and acceleration, verify safety margins and quantify strengthening effectiveness. Tower preliminary FEMs were implemented [15] and then developed further [4]. Structural analyses have been carried out using two different models: a "general purpose" FEM (ADINA [2]) and a specific code for masonry structures (3Muri [14]), through a frame with macro-elements and equivalent horizontal floors.

5.1 ADINA model

The FEM has been implemented by a direct transformation of the CAD geometric survey drawings, taking into account structural irregularities and morphological details (fig. 6). A mesh density optimization balanced accuracy and calculation time; the best FEM is characterized by an intermediate topology (14567 nodes and 20585 3D solid elements). The masonry characteristics have been identified not only considering code prescriptions and investigation results, but also a fitting on the experimental frequencies (table 1). An intermediate Confidence Factor (1.20) has been selected, depending on the construction knowledge level (ranging from 1.00, exhaustive, to 1.35, limited). The "Importance Factors" γ_I, used for load combinations to increase the seismic demand in the case of buildings of particular importance, are fixed in the seismic code [16] but redefined for cultural heritage in [17], according to artistic relevance and final destination [13]. In our case, γ_I has been taken equal to 1.0. For the limit state analyses, the elastic and design spectra definition followed again the code requirements (see tables 2–3 and fig. 7). In masonry aggregates, the reciprocal interactions must be carefully evaluated; therefore, constraint and contour conditions have been calibrated, generating the following tower base-fixed FEMs (fig. 8): (i) non-interacting with adjacent buildings; (ii) interacting with adjacent buildings, with rigid constraints in correspondence of the contact walls; (iii) interacting with adjacent buildings, simulating a portion of the contact walls with more flexible constraints, in order to consider the structural deformability; (iv) equal to the latter, but adding constraints simulating steel ties at the roof level, to inhibit relative displacements at the nodes. Several sets of calculations have been carried out (linear static analysis under vertical and horizontal loads, modal and response spectrum analysis).

Table 1: Material parameters used in the FEMs.

(a) code prescriptions					
material	E [N/mm^2]	G [N/mm^2]	density [kN/m^3]	f_m [N/cm^2]	shear resistance [N/cm^2]
ordinary masonry	1230	205	20	132.50	4.30
(b) parameters after the fitting on the experimental frequencies					
E=8000 N/mm^2			density=16kN/m^3		

Table 2: Italian seismic code expressions for the elastic response spectra.

horizontal acceleration								
$0 \leq T < T_B$	$S_e(T) = a_g \cdot S \cdot \{1 + T/T_B \cdot (\eta \cdot \beta_0 - 1)\}$							
$T_B \leq T < T_C$	$S_e(T) = a_g \cdot S \cdot \eta \cdot \beta_0$							
$T_C \leq T < T_D$	$S_e(T) = a_g \cdot S \cdot \eta \cdot \beta_0 \cdot (T_C/T)$							
$T_D \leq T$	$S_e(T) = a_g \cdot S \cdot \eta \cdot \beta_0 \cdot (T_C T_D/T^2)$							
soil category	a_g	S	T_B	T_C	T_D	β_0	η	ξ
A (rock)	0.25	1.00	0.15	0.40	2.00	2.50	1.00	5.00
vertical acceleration								
$0 \leq T < T_B$	$S_{ve}(T) = 0.9 \cdot a_g \cdot S \cdot \{1 + T/T_B \cdot (\eta \cdot \beta_0 - 1)\}$							
$T_B \leq T < T_C$	$S_{ve}(T) = 0.9 \cdot a_g \cdot S \cdot \eta \cdot \beta_0$							
$T_C \leq T < T_D$	$S_{ve}(T) = 0.9 \cdot a_g \cdot S \cdot \eta \cdot \beta_0 \cdot (T_C/T)$							
$T_D \leq T$	$S_{ve}(T) = 0.9 \cdot a_g \cdot S \cdot \eta \cdot \beta_0 \cdot (T_C T_D/T^2)$							
soil category	a_g	S	T_B	T_C	T_D	β_0	η	ξ
A (rock)		1.00	0.05	0.15	1.00	3.00	1.00	5.00

Table 3: Italian seismic code values for the design response spectra.

Confidence Factor CF (medium-high knowledge level)		1.2
Importance Factor γ_I		1.0
ordinary masonry buildings not regular in height		$q = q_0\ \alpha_u/\alpha_1$
$q_0 = 1.5$	$\alpha_u/\alpha_1 = 1.8$	$q_0 = 1.5$
Structure Factor for the Ultimate Limit State q_{ULS}		$q = 2.7$
Structure Factor for the Damage Limit State q_{DLS}		2.5

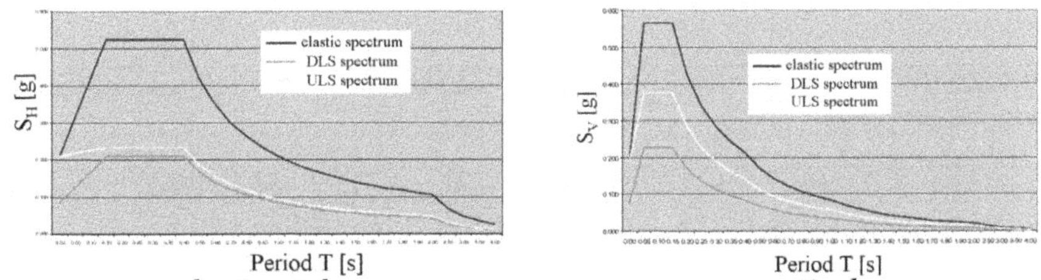

horizontal *vertical*

Figure 7: Elastic and design spectra.

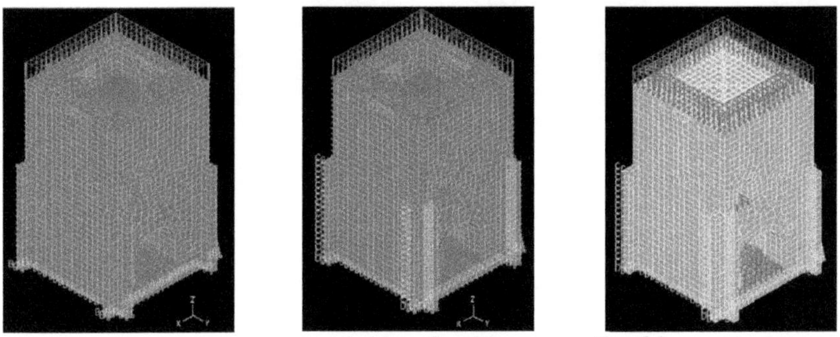

no constraints *rigid or flexible constraints* *flexible constraints and steel ties at the roof level*

Figure 8: FEMs with different constraint and contour conditions.

Table 4: Displacement values for linear static analysis.

linear static analysis displacement under vertical loads FEM (iii) [mm]	u	x	y	z
max.	1.70	0.50	0.30	0.00
min.	0.00	0.00	0.00	1.70
linear static analysis displacement under horizontal loads FEM (iii) [mm] – x direction	u	x	y	z
max.	6.70	6.37	3.68	0.60
min.			3.03	-3.07
linear static analysis displacement under horizontal loads FEM (iv) [mm] – y direction	u	x	y	z
max.	6.77	0.89	6.01	0.78
min.		-2.92		-3.19

Table 4 shows the calculated displacements for linear elastic analysis: in case of horizontal loads, the force distribution has been taken first mode proportional. Table 5 shows the first ten frequencies. Although a linear elastic behaviour (which is non-realistic) has been assumed in the calculations for the masonry, nevertheless the results provide interesting information on deformation and damage mechanisms. The earthquake damage subjected by the tower seems to be well described by the calculated modal shapes (modes 1 and 2 bending; mode 3 torsion), fitted with the experimental data (fig. 9 and table 5).

Table 5: Natural frequencies obtained for FEMs (iii) and (iv).

Natural frequencies	(iii)	(iv)	exp.	Natural frequencies	(iii)	(iv)	exp.
		[Hz]				[Hz]	
mode 01	4.810	4.811	4.500	mode 06	12.061	12.067	-
mode 02	4.927	4.930	5.700	mode 07	12.520	13.605	-
mode 03	8.673	8.674	7.500	mode 08	13.602	16.529	-
mode 04	10.855	10.868	-	mode 09	16.356	16.620	-
mode 05	11.700	11.704	-	mode 10	16.439	18.809	-

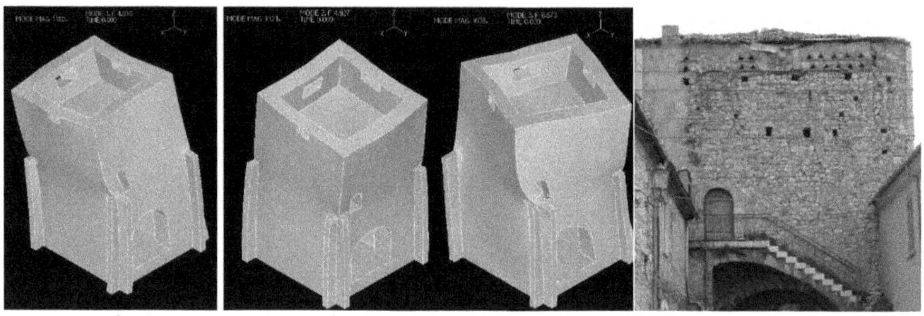

mode 1 *mode 2* *mode 3* *earthquake damage*

Figure 9: Calculated modal shapes and tower earthquake damage.

5.2 3Muri model

Also in this case, several models of the tower have been implemented: base-fixed with (j) and without (jj) interactions to adjacent buildings; base-fixed with (jjj) and without (jv) masonry improvement through mortar injections. The tower vertical components have been schematized through macro-elements (rigid nodes, openings, masonry belts, vertical walls), while the horizontal ones described by floor elements with infinite axial stiffness and no bending.

Push-over analyses have been performed, taking into account 24 cases deriving from the load type combinations (mass proportional or first mode proportional), direction of the seismic action and load eccentricity, with the aim of identifying the most dangerous scenario. For each load case considered, safety margins have been verified, as requested by the Italian seismic code. In models (j) and (jj), a sudden collapse happens at the building base; it is due to the great stiffness of the masonry walls at this level, as showed by the push-over curve, missing the descending branch (fig. 10). On the contrary, in models (jjj) and (jv), damage happens in the masonry walls placed over the constraints representing the confining effect due to the presence of adjacent buildings. These walls being thinner than the basement, the non-sudden collapse derives from the macro-elements plasticization (fig. 11). It is worth noting that in models (jjj) and (jv) the code required performance is verified, coherently with the features of the push-over analyses. These analyses can well represent the global behaviour of the building under the effect of horizontal actions, but local damage due to shaking effects can be lost. As a matter of fact, the Italian code foresees also the analysis of local damage mechanisms, regarding out-of-plane collapse of single panels and overturning of entire vertical walls (if monolithic), as shown by many earthquake damage patterns. In conclusion, the structural designer must know the different capabilities and limits of the various methods.

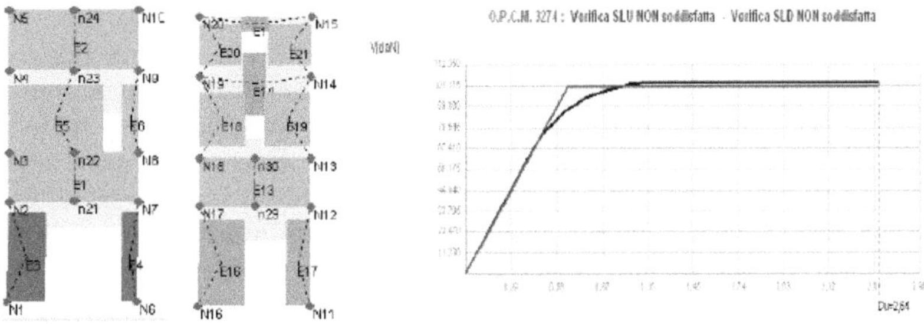

Figure 10: Models (j) and (jj), collapse mechanism and push-over curve.

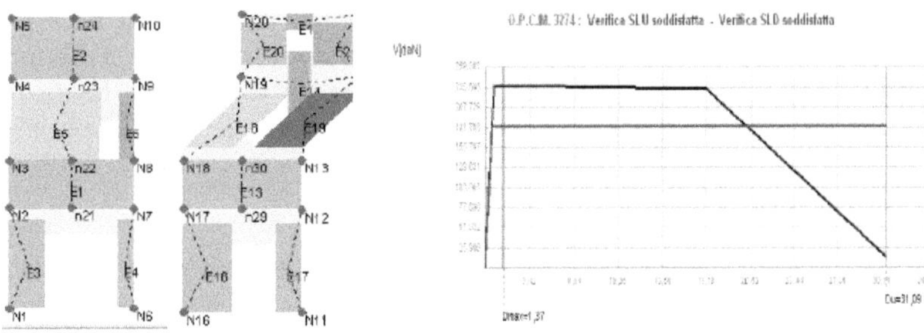

Figure 11: Models (jjj) and (jv), collapse mechanism and push-over curve.

Figure 12: Tower roof strengthening.

6 The rehabilitation design of Marchesale Castle

Thanks to experimental/numerical data, an antiseismic improvement proposal has been suggested, to harmonize protection and conservation, in agreement with the codes. Vertical walls strengthening (and their reciprocal interconnections) is recommended, through the insertion of steel-ties at each level (using, if possible, innovative materials as shape memory alloy devices [1]), in both horizontal directions, together with a limited reconstruction of the most damaged masonry portions by using similar and compatible materials, avoiding local stiffening. Injections of compatible mortar, where necessary, have to be done after a detailed identification of mixtures and procedures. Only if indispensable, the "a sacco" masonry can be strengthened by the insertion of transversal connections, with the aim of linking the external wall sheets with each other. Vaults reinforcing has to be carried out by using conventional methods (repairing and thickening by layers of thin bricks), but rejecting r.c. coping. All the floors should be light enough, preferably wooden made, but provided by adequate stiffness (using cross-bracing systems and additional wood planking), anchoring them carefully to the vertical walls, keeping off the use of floor r.c. string-courses and preferring steel string-courses; at the roof level (fig. 12), a reinforced masonry-course (or reticular frame steel string-course or wood string-course) can be provided, together with effective connections of the wooden roof components.

7 Conclusions

The work suggests strengthening interventions for the Marchesale Castle old masonry tower, damaged by the 2003 earthquake, after discussing experimental and numerical results. It is possible to give the structure its original stiffness and place resistant systems capable of improving the building dynamic response to seismic shear-force actions (according to the existing set of rules), respecting the original structure as much as possible, and avoiding conflicts between conservation criteria and antiseismic requirements.

References

[1] Auricchio, F., Faravelli, L., Magonette, G. & Torra, V., (eds). *Shape Memory Alloys. Advances in Modelling and Applications,* CIMNE: Barcellona, 2001.

[2] Bathe K.J., *ADINA*, http://www.adina.com/, 1986.

[3] Buffarini, G., Clemente, P. & Paciello, A., Rilievo delle vibrazioni ambientali di Palazzo Marchesale a San Giuliano di Puglia (CB). ENEA Report, 2006.

[4] Cami R., Tecniche antisismiche moderne e metodi di calcolo strutturale per la protezione dei beni storico monumentali: il caso della torre di Palazzo Marchesale a San Giuliano di Puglia, Ph.D. Thesis, University of Ferrara, Faculty of Engineering, 2007.

[5] Carpani, B. & Indirli, M., Commenti sul progetto esecutivo di consolidamento statico e di riqualificazione funzionale del Palazzo Marchesale situato a San Giuliano di Puglia (CB). *ENEA Report*, 2005.

[6] Doglioni, F., Moretti, A. & Petrini, V., *Le Chiese e il Terremoto*, Lint: Trieste, 1994.

[7] Doglioni F., Codice di Pratica per la progettazione degli interventi di riparazione, miglioramento sismico e restauro dei Beni Architettonici danneggiati dal terremoto umbro-marchigiano del 1997. *Bollettino Ufficiale Regione Marche* n. 15, 2000.

[8] Giuffrè A., *Monumenti e Terremoti*. Multigrafica Editrice: Roma, 1988.

[9] Giuffrè A., *Sicurezza e Conservazione dei Centri Storici. Il Caso di Ortigia*. Laterza: Bari, 1993.

[10] Indirli, M., Clemente, P. & Spadoni, B., The reconstruction of San Giuliano di Puglia after the October 31st 2002 earthquake. *Proc. 13th World Conference on Earthquake Engineering*, August 2004, Vancouver, B.C., Canada.

[11] Indirli, M., Clemente, P., Spadoni, B., Cami, R., Speranza, E., Mucciarella, M. & Pistola, F., Seismic protection of historical centers using innovative techniques, with focus on San Giuliano di Puglia after the 2002 Molise earthquake. *Proc. IV SAHC International Conference,* November 2004, Padua, Italy.

[12] Indirli, M., Cami, R., Carpani, B., Algeri, C., Panzeri, P., Rossi, G. & Piova, L., The antiseismic rehabilitation of Marchesale Castle at San Giuliano di Puglia, *Structural Analysis of Historical Constructions (SAHC)*, eds. P.B. Lourenço, P. Roca, C. Modena & S. Agrawal, New Delhi, India, 2006.

[13] Modena, C., Casarin, F., Valluzzi, M.R. & Da Porto, F., Codes of practice for architectural heritage in seismic zones. *Structural Analysis of Historical Constructions (SAHC)*, eds. P.B. Lourenço, P. Roca, C. Modena & S. Agrawal, New Delhi, India, 2006.

[14] Lagomarsino, S., Penna, A. & Galasco A., 3Muri, *Linear and non-linear static analysis of masonry and mixed buildings*, http://www.stadata.com, 2005.

[15] Pistola F., Protezione dai terremoti con tecniche innovative, con particolare riferimento al patrimonio culturale: il caso di San Giuliano di Puglia, Degree Thesis, University of Florence, Faculty of Architecture, 2005.

[16] Presidenza del Consiglio dei Ministri, OPCM 3274, 3316, 3431, *Gazzetta Ufficiale*: May 8th, 2003, Ser. Gen. n. 105, Suppl. Ord. n. 72; October 10th, 2003, Ser. Gen. n. 236; May 10th, 2005, Ser. Gen. n. 107, Suppl. Ord. n. 85.

[17] Presidenza del Consiglio dei Ministri, *Linee Guida per l'applicazione al patrimonio culturale della normativa tecnica*, 2006.

Seismic retrofitting of adobe masonry buildings based on collapse analysis

T. Bakeer & W. Jäger
Dresden University of Technology, Germany

Abstract

A novel approach is presented for retrofitting and reconstruction of cultural heritage buildings based on collapse analysis. The computer modelling is carried out through LS-DYNA code, where the combined finite/discrete element method is adopted. The approach is applied for two rooms in Sistan's house in Bam citadel, which collapsed under the earthquake of 2003. The proposed methodology is used to verify the retrofitting measures that were implemented into the structure to resist the forthcoming earthquakes. Further experimental investigations are performed to enhance the performance of adobe material under earthquake actions and the determination of the suitable reinforcement materials is achieved.

Keywords: adobe masonry, collapse analysis, macro modelling, combined FEM/DEM, explicit dynamics.

1 Introduction

Bam Citadel (Arg-e Bam) is one of the largest adobe complexes in the world and plays an important role in the history of Iran. The citadel which is traced back to the Achaemenid period (6th to 4th centuries BC), and is considered the most representative example of a fortified medieval construction built by adobe masonry. The citadel is situated atop a hill in the northwest old city of Bam on an area of 180,000 square meters. Bam Citadel was destroyed during an earthquake in December 2003. Immediately after the disaster, activities were started to preserve the destroyed monument site and to clear the question how to deal with the remains. UNESCO included the Citadel with its unique surroundings and landscape in the list of World heritage sites under risk. Now the consolidation, restoration and reconstruction are under progress.

Several efforts were made by the Iranian authorities and the international community to reconstruct the collapsed citadel. Within the framework of reconstruction of the Bam citadel by international experts, our German team undertook a part in Sistani's House. In order to develop a strengthening methodology for Sistani's House, two rooms R 0.11 and R 0.12 in the northwest corner were selected as the pilot project (fig. 1).

Figure 1: Sistani's House, (a) the ground plan, the selected part for the pilot project marked in red (b) 3D model of the house, Einifar [1], (c) Sistani's House after the earthquake of 2003; the debris inside the rooms is removed.

2 Investigations of the damage and reasons for collapse

Several processes are carried out in situ, starting with the removal of the debris and survey documentation of the ruins. The key question to be answered is the allocation of the fallen debris to its original built-in position. Following that, it becomes possible to analyze the process of the collapse in order to verify the vulnerability of the structure.

The survey was performed by a total station and photogrammetry. The drawing generated from the photogrammetric images served as a basis for recording the situation of the damage; it also gives essential information about the original form of the house. The damage patterns show a large diversity and range from capillary cracks to large gaps and debonding along different building phases.

The remains and the damage patterns observed in situ indicate that the collapse initiated due to out of plane failure of the walls. Such failure could be a result of the vaulted roofing system which has high vulnerability to collapse under horizontal earthquakes.

3 Principles of dealing with remains and reconstruction

The enormous size of the destruction in the citadel is associated with the weak brittle material of adobe masonry. It is difficult in such situation to find the appropriate solution for the restoration, reconstruction and retrofitting of the highly damaged architectural heritage. However, some guidelines that match the special demands of modern UNESCO world cultural heritage site are adopted for the reconstruction process.

At first, it is essential to retrieve as much as possible the remains of historical substance to reinstall them as effective part for later reconstruction phase.

The behaviour of brittle material such as adobe masonry could be enhanced by increasing the ductility which offers the ability to prevent sudden collapse. The presence of a vaulted roofing system is the reason of the overall stability of the structure being lost during the earthquake. Therefore, the implementation of ring reinforcement is of high importance.

4 Collapse analyses as tools for verification

In order to explore the performance of the structure which collapsed under the earthquake of 2003, a collapse analysis is carried out by means of LS-DYNA code [6].

The elements of the structure (thick walls, vaults and piers) comprise a large number of bricks. Therefore, the generation of the geometry brick by brick will be highly time-consuming; furthermore, the geometry of the bricks is not known beforehand. As a result, micro-modelling cannot be applied in such a case. An alternative approach that can overcome the problem is to employ macro-modelling, which needs a small modelling effort for large structures. It reduces the model size and calculation time as well.

In adobe masonry, mud mortar has been used to assemble the mud bricks. This points toward getting material continuity in between the brick units and mortar as well as approximately similar mechanical characteristic for masonry in orthogonal directions.

Since the structure is already destroyed by the earthquake of 2003, the definition of geometry is based on the available data in situ, that is, the remaining parts, and the available pictures before the collapse. The whole geometry of the structure is divided into discrete elements by means of CAD tools (fig. 2).

Each pair of two adjacent discrete elements is sharing a contact interface which ties both interface elements prior to the failure. This tends to generate separation of the tied discrete elements when failure of contact takes place. Those interfaces are representing the locations of potential cracks. Without those

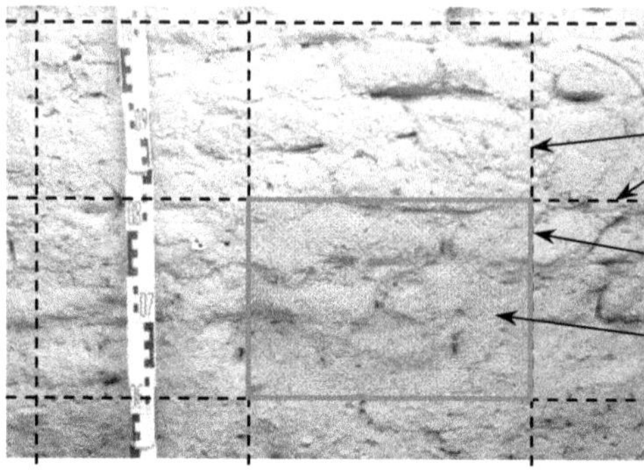

Figure 2: Discretization of adobe masonry wall.

interfaces the calculation process terminates, due to large deformations of finite elements at failure. LS- DYNA tiebreak contact model has been used to model the planes of failure. Tiebreak contact is active for nodes which are initially in contact. The slave nodes are sticking permanently until reaching the failure criterion. After failure, this contact option behaves as frictional contact.

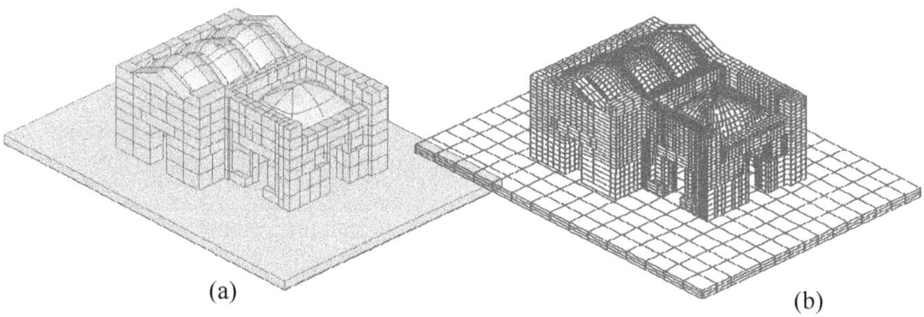

Figure 3: (a) Discretized geometry (b) finite element mesh.

The discretization is performed in a regular manner so that regular geometrical shapes are produced and acute angles are avoided; the later cause problems with the handling of contact (fig. 3(a)). In addition, mapped meshing is applied comprising eight-node brick finite elements with a single integration point (fig 3(b)). This prevents the negative volume problem at an early stage, which terminates the calculation in explicit solvers.

The LS-DYNA soil and foam constitutive model has been employed for modeling the material of adobe masonry. The material properties used in our model are based on values adopted by Taheri [3], Kiyono and Kalantari [4] and in [2].

The collapse analysis of the structure shows that the collapse initiates in the longitudinal direction of the earthquake; this can be explained by the high acceleration at the beginning of earthquake (fig. 4). A few seconds later, the

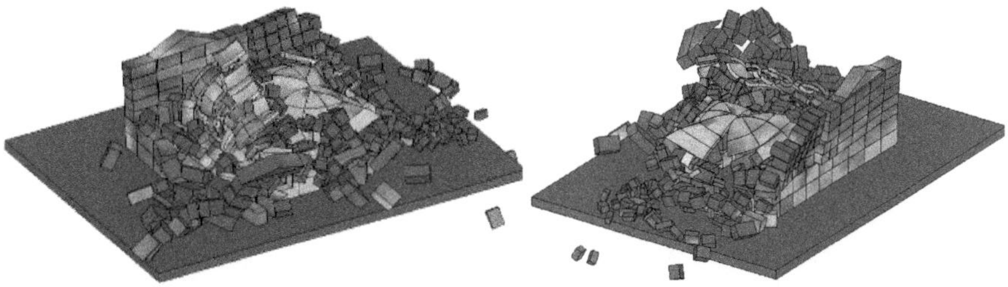

Figure 4: Collapse state of the structure at time $t = 18.80$ s.

walls perpendicular to the transversal earthquake direction are collapsed. At this time, there is an increase of earthquake intensities along transversal direction which can explain this collapse. The order of collapse sequence as noted in collapse analysis is identical to that predicted by the investigations on the ground. The falling debris on the ground show clearly that the direction of the collapse is mainly along east-west direction. At the last stage of the analysis, out of plane failure of the walls is the main failure mode, which fits with results described by Kiyono and Kalantari [4], for adobe masonry buildings in Bam. The relatively high value of friction coefficient 0.54–0.62 causes this mode of collapse.

5 Reinforcement of the building

The results of collapse analysis have showed that collapse was mainly initiated due to out of plane failure of the walls. This kind of failure is possible in such structures due to vault-form construction of the roofs. Therefore, the capacity of the structure can be enhanced to resist out of plane actions by implementing ring reinforcement. The vertical reinforcement enhances the integrity and the load bearing capacity as well.

Fibreglass nets with clay-cement grout are supposed to be used as reinforcement; a series of pull-out tests have been carried out in order to determine the proper reinforcement and the grout material to be used with adobe masonry [2]. The reinforcement was built in 36 cm into the wall, which is the tested anchor length. For a standard fibreglass net built into a borehole $d = 30$ mm, the average maximum pull-out force for clay-cement grout was 9 kN (fig. 5).

In order to simulate the pull-out failure of reinforcement, a breakable bar linked to masonry elements via nodes along the reinforcement locations have been used. Although LS-DYNA provides various possibilities for modelling the reinforcement, each has its own drawback. However the spotweld linkage has been defined along a string of nodes that belong to each reinforced discrete element. The distance between the pair of nodes linked by spotweld is 0.5 m.

The calculation of resisting reinforcement has been performed in iterative scheme. Several reinforcement trials have been tested by collapse analysis on Room 11. Fig. 6 shows the reinforcement arrangement (horizontal ring reinforcement, vertical reinforcement, and vault reinforcement) that was used for calculation trials. The spotweld failure values of the reinforcement are given in table 1.

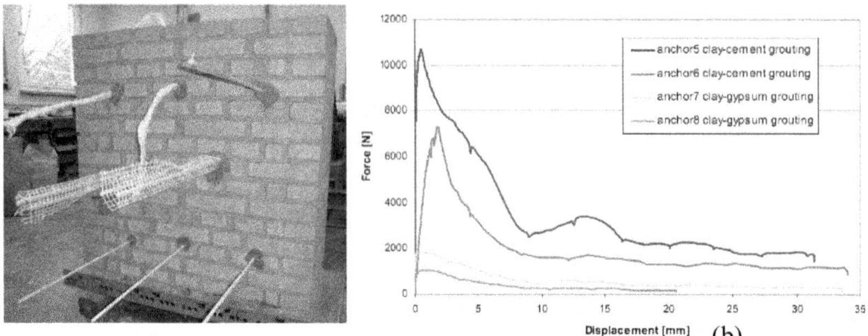

(a) (b)

Figure 5: Pull out test of reinforcement (a) test setup, (b) pull-out force versus displacement plot for different anchoring types [2].

Table 1: Reinforcement bonding capacity.

Trial No.	Maximum bonding force (kN/50 cm)		
	Horizontal Reinf.	Vertical Reinf.	Vault Reinf.
#1	10	10	-
#2	50	10	-
#3	100	25	-
#4	100	25	25

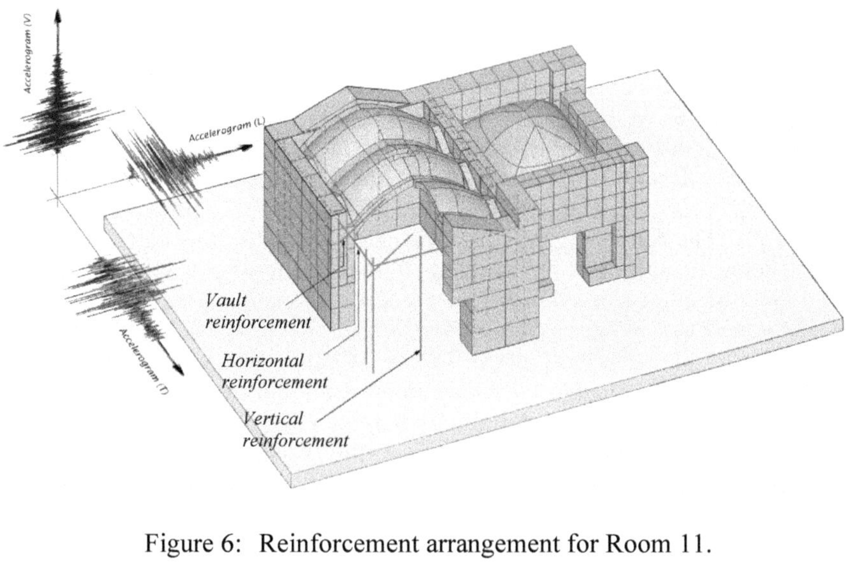

Figure 6: Reinforcement arrangement for Room 11.

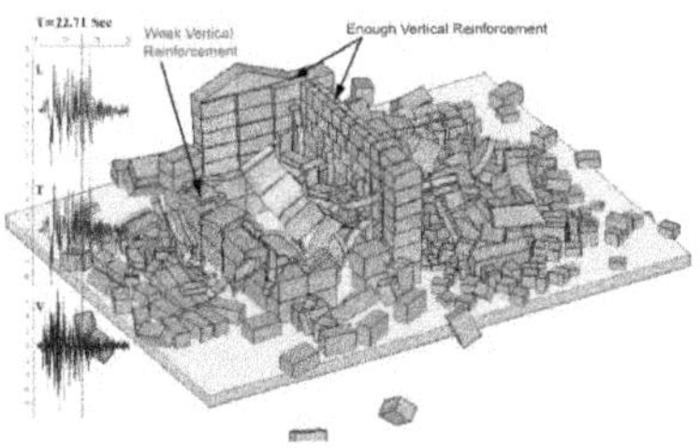

Trial #1, collapse state at T=17.89 Sec

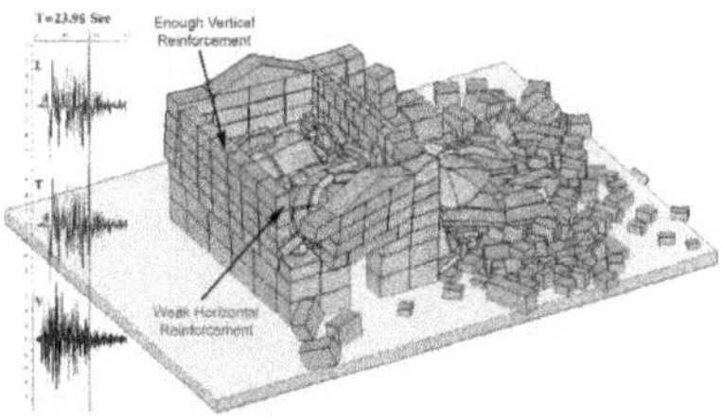

Trial #2, collapse state at T=22.71 Sec

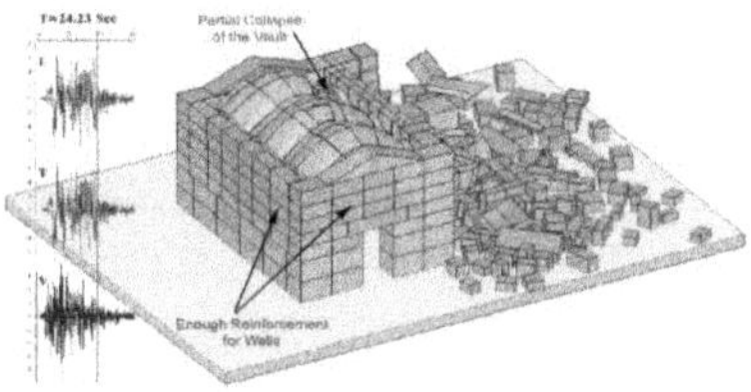

Trial #3, collapse state at T=19.61 Sec

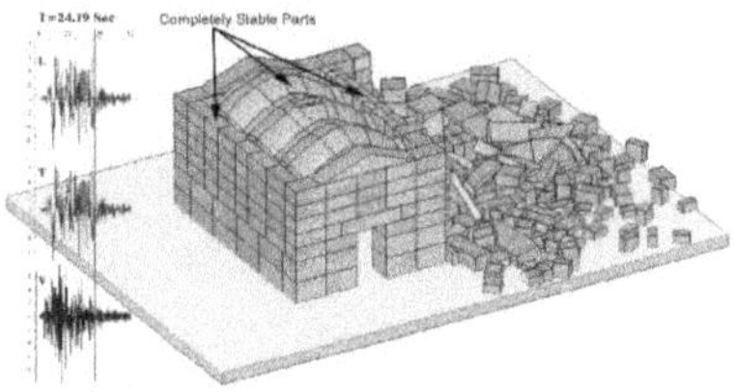

Trial #4, collapse state at T=24.19 Sec

Figure 7: Collapse states at different reinforcements' trials.

In the 1st trial, vertical and horizontal reinforcement were added to the walls; the final remains showed that some walls remained intact at the end of earthquake and others collapsed partially and the vault collapsed due to progressive collapse. The reinforcement was sufficient for some walls, but must be increased for others. We have to consider that the creation of new strong parts by reinforcement may change the collapse mechanism. In the 2nd trial, the horizontal ring reinforcement was increased, but that was not sufficient and the ring reinforcement failed. In the 3rd trial, the walls of the room showed good stability but the vaults partially collapsed and showed large deformations. Thus, one task that should be achieved as a next step is to add reinforcement to the vaults (fig. 7).

6 Improving the ductility of units

To improve the ductility of adobe material, adobe units are produced with sisal, coconut, flax, hemp and palm fibres and a series of experiments were carried out to decide the suitable material of reinforcement.

The tests were performed in a Dresden laboratory as well as in Bam, in order to obtain results under real conditions. The results led to the recommendation of the use the ratios of clay and sand (70: 30) and fibres (0.6% of the weight of the sand-soil mix). The tests showed no increase in compressive and tensile strength. The stress-displacements relationship from the compression and tension tests showed considerable amount of deformation prior to failure. This indicates an enhancement in the ductility of the material of the units (fig. 8).

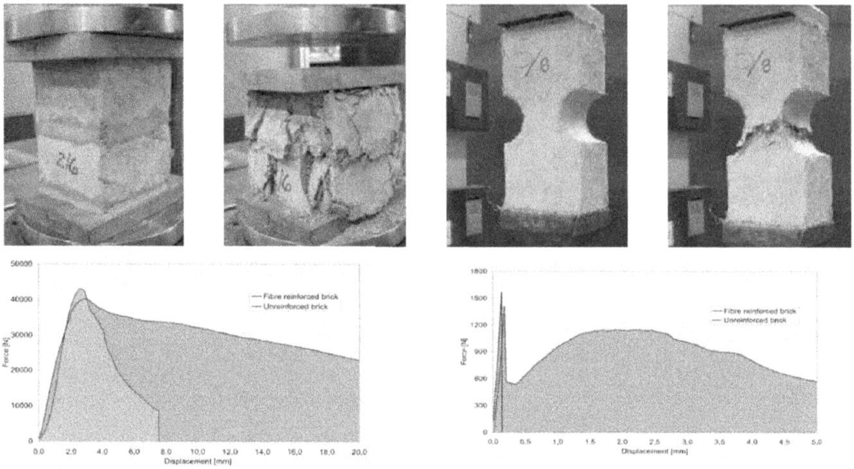

Figure 8: Laboratory tests showing higher compression and tensile strength for a palm fibre reinforced adobe.

7 Implementation of the reinforcement

The remaining walls are repaired first with the fibre reinforcement and the walls are brought up to an equal level by removing remnants of unsustainable masonry. Following that, vertical anchors of glass fibre are inserted as described by Jäger and Fuchs [5].

Scaffolding, upon which the drilling machine for vertical boreholes can be based, is installed next to the walls. Boreholes are drilled at an interval of 1 m and reach down to a depth of 1.5 m below the ground level. The vertical anchor consists of three fibre rods with a diameter of 8 mm. At its lower end, the bundle of three rods is spliced up. It is then inserted into a block of fine concrete that ties the vertical anchor to the ground. Horizontal reinforcement layers of fibreglass mesh are added, each at 0.5 m height. A horizontal rode of fibreglass is fixed to the vertical anchors as well. The ring beam reinforcement of the vaults consists of six fibreglass rods (width 0.6 m, height 0.3 m). For the vaults, the transversal arches are wrapped with the same fibreglass mesh that is used for horizontal joints and covered with traditional clay plaster (fig. 9).

8 Summary and outlook

A novel approach is presented for retrofitting and reconstruction of cultural heritage buildings based on collapse analysis. The approach is applied for two rooms in Sistan's house in the Bam citadel which collapsed under the earthquake of 2003. The proposed methodology used to verify the retrofitting measures that were implemented into the structure to resist the forthcoming earthquakes.

Figure 9: Reconstruction and implementation of reinforcement [5].

Acknowledgments

The authors would like to thank UNESCO, the Japan Funds-in-Trust for World Cultural Heritage and the Federal Foreign Office of Germany for their generous funding of the project. Furthermore, thanks are also due to the Recovery Project of Bam's Cultural Heritage and its efficient and helpful staff.

References

[1] Einifar A., *3-Dimensional Reconstruction and Virtual Reality of Bam and its Cultural Landscape*, NII-ICHTO University of Tehran-Razahang, University Collage of Fine Arts – UT, PowerPoint presentation, 2007.

[2] Jäger, W., (ed). *Interim Report for Rehabilitation Concept for Sistani's House-Bam Citadel.* Report No.40632, Jäger Consulting Engineers and Dresden University of Technology, Dresden, Germany 2007.

[3] Taheri, S.M., *Finite Element Modeling of Adobe Masonry Buildings under Earthquake Loads and Contribution of Reinforcement in Enhancement of Load Bearing Capacity*. MSc Thesis, Dresden University of Technology: Dresden, Germany 2006.

[4] Kiyono, J. & Kalantari, A., Collapse mechanism of adobe and masonry structures during the 2003 Iran Bam earthquake. *Bulletin of the Earthquake Research Institute, University of Tokyo*, **79(3/4)**, pp. 157-161, 2004.

[5] Jäger, W. & Fuchs, C., Restoring adobe masonry on a World Cultural Heritage site: The Sistani House in Bam, Iran. *Proc. of the 8th International Seminar on Structural Masonry*, Istanbul, 2008.
[6] Hallquist, J.O., *LS-DYNA Theory Manual*, Livemore Software Technology Corporation: California, USA, 2005.

Strengthening a heritage structure with self compacting concrete: an experimental study

N. Torunbalci, I. Ediz & F. Sutcu
Faculty of Architecture, Istanbul Technical University, Istanbul, Turkey

Abstract

Heritage structures consisting of masonry walls are under high risk in the case of a severe earthquake. This study experimentally investigates the behaviour of brick masonry wall samples strengthened by self-compacting concrete and standard concrete on one side and includes a key study representing an example of application to historical structures. Comparisons are made between the change in the load-bearing capacities and behaviour of the specimens following the strengthening. It is determined that the improvements in the specimens become much more pronounced when the self compacting concrete rather than the conventional concrete strengthening is applied.
Keywords: masonry, self-compacting concrete, brick masonry, one side retrofitting.

1 Introduction

The majority of buildings in Turkey are masonry structures and the masonry walls are mostly made of bricks. In addition to numerous structures of monumental and historical value, the majority of the structures constructed both in rural and urban areas are of the masonry structure type. Although masonry brick walls are known to have high vertical load bearing capacity, they are weak against lateral loading such as earthquakes. In this context, it is crucial to have the existing masonry structures protected against the earthquakes by increasing their strength.

The damage to the masonry structures is attributable to inferior soil conditions, non-conformity with the requirements of relevant codes, defective design and application as well as wear and tear. As a result, earthquakes cause major damage to the structure or else, result in the collapse of the structure as a whole. It is therefore necessary to strengthen buildings with a weak structural system and structures already damaged in past earthquakes before major earthquakes. The most popular method used is the strengthening of the existing wall by using shotcrete or conventional concrete with steel wire mesh reinforcement.

However, many problems are usually experienced in practice. Placing and compacting of the concrete cannot be carried out effectively, due to the narrow formwork. Increasing the thickness of the concrete layer can eliminate the problem; however this in turn gives rise to undesirable loss of the footing area. Moreover, the pressure of shotcrete application might cause major damage on a historical masonry wall. Such type of strengthening is therefore not preferred, even if it greatly contributes to the structural performance. If self-compacting concrete is used, the problems in question may be solved effectively. Based on the foregoing arguments, experimental investigations have been conducted in this study [1] on masonry structures. Unreinforced masonry walls (URM) are used to simulate the masonry structural walls of an historical structure, and they are strengthened with the self-compacting concrete and conventional concrete with or without steel wire mesh reinforcement. A comparison of improvements was made with reference to the load-bearing and deformation capacities.

Many studies have been carried out concerning the methods used for the strengthening of masonry walls. On the other hand, it can be seen, through a review of the relevant literature, that a very limited number of studies are available concerning the application of conventional concrete with the steel wire mesh reinforcement. Furthermore, no studies could be found on strengthening by contemporary concrete technology, i.e., self-compacting concrete.

Shing *et al.* [2] investigated the influence of the horizontal and vertical strengthening ratio of the strengthened walls' ductility, energy dissipation capacities, stresses and lateral resistances. Conducted on concrete masonry and brick samples, the study incorporates hollow blocks filled with mortar and strengthened with horizontal and vertical reinforcement. It has been found out that, vertical strengthening increases the bending strength of the wall and that the diagonal cracks formed as a result of the shear forces depend on the number of the steel strips used horizontally and vertically and that increasing the ratio of the reinforcement increases the ductility energy dissipation capacity.

Experiments were carried out on samples with and without reinforcement in another study by Drysdale and Khattab [3] to investigate the in-plane behaviour of concrete masonry walls that are plastered, strengthened and subjected to biaxial loading. Sample walls were constructed with strengthening elements placed at various angles depending on the loading setup and joints that are parallel and orthogonal to the strengthening joints. Experimental work on the brick wall samples without reinforcement reveals that failures are observed in the form of vertical cracks when loading at 0° and 90° with respect to the joints, whereas bad joint failures are observed during inclined loading at 45° to the joints. In loading reinforced samples, however, vertical cracks formed at 0° and 45°, whereas at 67.5°, the failures are due to the shears of the joints. It is observed that strengthening insures considerable improvement in the deformations.

The behaviour of typical residences constructed in rural areas were investigated in a research carried out by Agarwal and Thakkar [4] on half scale brick masonry buildings under cyclic loading and looking at the improvement obtained by strengthening via reinforced concrete bonds. The strengthening is provided on the vertical r/c bonds in combination with lintel bands and sill-bands that are associated with them. In repair works, steel wire mesh reinforcements, which are plastered by cement and epoxy based mortar, are installed. It was found that lintel bands and sill-bands are effective in minimizing the cracks at the bottom. On cracked portions, the steel wire mesh reinforcements accompany epoxy based mortar. This contributes to the ductility and deformation capacities. Where the steel wire mesh reinforcement and cement base mortar are used, however, the models undergo brittle failure; however they exhibit a high level of strength.

2 Purpose of experimental work and test plan

In this study conventional brick walls are strengthened using self compacting concrete (SCC) and conventional concrete (CC) on one side and the aim is to determine the effect of this strengthening on the behaviour and the strength of the walls by an experimental approach. The brick wall samples were mounted on the testing rig in such a way so that the bed joints formed an angle of 45°with the direction of vertical loading. Loading is exerted onto concrete shoes constructed on opposing corners of the brick masonry specimens [1]. Half of the specimens were pre-damaged and repaired with epoxy resin before strengthening and comparisons were made with reference to the change in the load-bearing capacity and behaviour following the strengthening. Within the scope of this study, the objective was also to make a comparative analysis on the effects of different types of concrete and inspect the displacements and failure modes, for which a relevant specimen series were produced, corresponding to the resulting alternatives of respective parameters.

Eighteen brick wall samples about 185×185 mm in size are used in the experiments. Brick wall samples are erected for the experiment by using bricks in standard sizes of 190 mm × 90 mm × 50 mm, reduced to the size 95 mm × 45 mm × 25 mm. Half of each series are subjected to pre-damaging process almost up to the level of collapse. These damaged specimens are then repaired with epoxy resin as seen in fig. 1. Together with non-damaged specimens, these pre-damaged-and-epoxy-repaired specimens are then loaded until the occurrence of failure. The behaviour of damaged specimens is expected to be different from that of non-damaged specimens. The aim here is to see the effects of steel mesh reinforcement on SCC strengthening.

A notation is adopted for the brick wall samples in accordance with the type of strengthening. Thus, D stands for the pre-damaged brick wall samples, ND for non-damaged samples, SCC for self-compacting concrete, CC for conventional concrete and Φ is added in case steel wire mesh reinforcement is used. URM stands for un-reinforced masonry wall and D-ER stands for the specimens which are damaged and then repaired with epoxy. Accordingly, ND-SCC stands for the non-damaged samples strengthened by self-compacting concrete and no steel reinforcement. Likewise, D-CC Φ stands for the pre-damaged samples strengthened by conventional concrete and steel reinforcement.

Figure 1: Damaged and epoxy-repaired wall specimen (D-ER).

2.1 Properties of materials incorporated

The joints of the brick wall samples are accepted as 5 mm thick and therefore, the average grain size of the sand is selected as 1 mm. The specific weight of the sand is 2650 kg/m^3. Type PC 32.5 and Type PC 52.5 grade cement are used in the mortars for erecting the brick wall samples and the top rows of the brickwork, respectively. The specific weight of the cement is 3125 kg/m^3.

The mortar used in erecting brick wall samples has a cement/sand ratio of 1:3.5 and the wet mortar consistency, as prescribed by the flow table testing, is 1.48. The specific weight of the mortar is 2130 kg/m^3. 9 samples 40×40×160 mm in size were prepared in order to determine the mortar strength and these samples were tested in groups of three on the 7th, 14th, and 28th days, respectively. The average mortar compressive strengths on the 7th, 14th, and 28th days are 8.33, 11.78 and 16.24 N/mm² respectively. In repairing the pre-damaged brick wall samples, epoxy resin is employed. This is a two-part, solvent free epoxy resin that can be used in humid environments. The 7days compressive strength of the epoxy resin used for repair is 75 MPa.

The steel wire mesh with a plain surface and a wire diameter of ø3, manufactured from S500 grade steel and braided with a grid spacing of

50×50 mm, is provided to supplement the strengthening process to strengthen reinforcement.

Self-compacting concrete has a spread diameter of 0.75m and specific weight of 2.350 kg/m^3. Compressing strength of the self-compacting concrete is tested on samples settled on the steel formworks sized 40x40x160mm. The samples are tested by the compressive testing apparatus capable of exerting 200 KN on the 7th, 14th, and 28th days following concrete settling. The respective compressive strengths are determined as 40.33, 45.38 and 56.39N/mm^2.

While determining the compressive strength of the conventional ready-mixed concrete with a maximum grain diameter of 12 mm, cubic samples 150x150x150mm in size are tested. The compressive strengths on the 7th, 14th, and 28th days are determined as 24.00, 29.18 and 38.80 N/mm^2, respectively.

2.2 Test schedule

Four series were strengthened with self-compacting concrete and 2 with conventional ready-mixed concrete. Apart from the 2 series strengthened with SCC, wire mesh reinforcement was used on all specimens fixed at about 10 mm in front of the surfaces of the brick wall samples so that they are located at the centre of the 20 mm thick concrete layer on surfaces of the samples as shown in fig. 2.

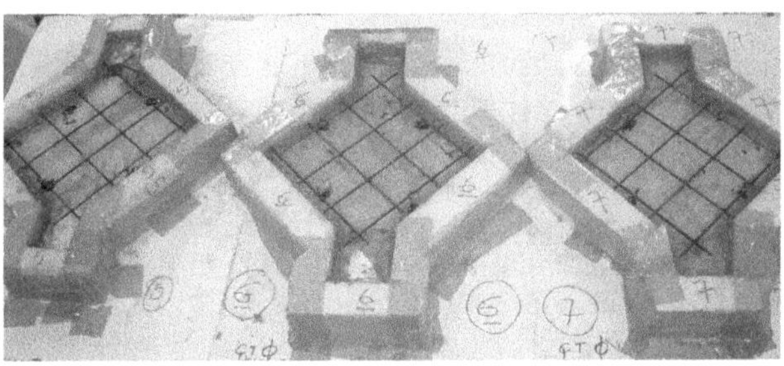

Figure 2: Fixed steel wire mesh reinforcements.

Half of each group were first subjected to loading until severe damage occurred. All cracks induced on the pre-damaged walls are repaired with epoxy resin.

Self-compacting concrete supplied from the concrete plant is applied on the surface of pre-damaged and non-damaged URM wall samples in layers 20 mm thick. The samples with strengthening provided by self-compacting concrete on one side only are kept in the formwork for 48 hours to get hardened.

In order to compare the self-compacting concrete strengthening with the conventional concrete strengthening, a total of 6 brick wall samples consisting of 3 pre-damaged and 3 non-damaged samples, are strengthened by conventional concrete on one side.

The brick wall samples were mounted on the testing rig in such a way so that the bed joints formed an angle of 45°with the direction of vertical loading. The crack propagation is observed and recorded by measuring the load-deformations via the mechanic strain gauges (fig. 3).

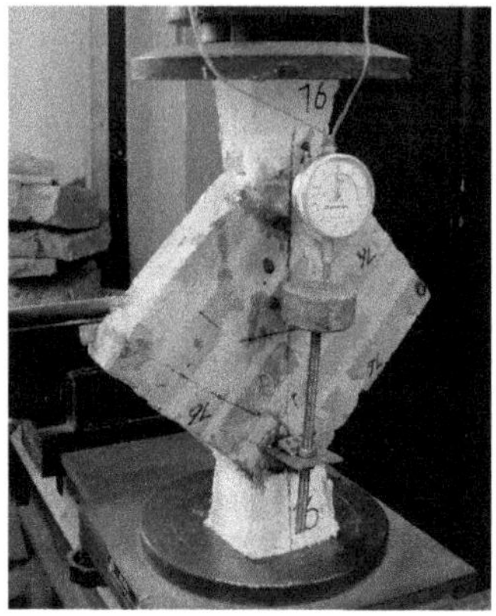

Figure 3: Experimental layout and brick wall specimens.

3 Results of experiments

This section provides the evaluation of the results of the experiments carried out on the brick wall samples. Experiments are performed on 6 different series.
In the course of determining, in particular, the behaviour of the brick wall specimens, 3 URM specimens are pre-damaged under axial compression and the cracks on the pre-damaged walls are repaired by the epoxy materials and subject to further loading. The maximum stress sustained by the specimens is calculated from eqn (1) in accordance with ASTM standards [5]:

$$S_S = \frac{0.707P}{A_n} \tag{1}$$

where S_S is the shear stress based on the net area, P is the applied load and A_n is the area of the specimen calculated according to eqn (2):

$$A_n = \left(\frac{w+h}{2}\right)tn \tag{2}$$

where w, h and t are specimen width, height and thickness, respectively and n is the percentage of the area of the unit that is solid, .as a decimal. Average maximum strengths of the URM walls subjected to loading and after repair by the epoxy materials following pre-damaging are determined as 1.53 MPa and 2.98 MPa, respectively. The results are shown in table 1, where COV represents the coefficient of variation for each series. It is determined that the displacements of the URM walls are increased twice as much following the epoxy repair. The failure loads of brick wall samples including URM are provided in fig. 4.

As indicated by the table and the figures, the strengths of brick wall samples as strengthened on one side vary depending on numerous parameters. The load bearing capacities of the walls strengthened on one side are to a great extent changed. On the other hand, the deformations are naturally decreased when additional strengthening is provided. The degree of decrease in the deformation stands at 10%.

The highest ultimate failure loading and stress capacities are observed on ND-SCCΦ brick wall samples, i.e., non-damaged samples strengthened by self compacting concrete with steel wire mesh reinforcement. The lowest failure

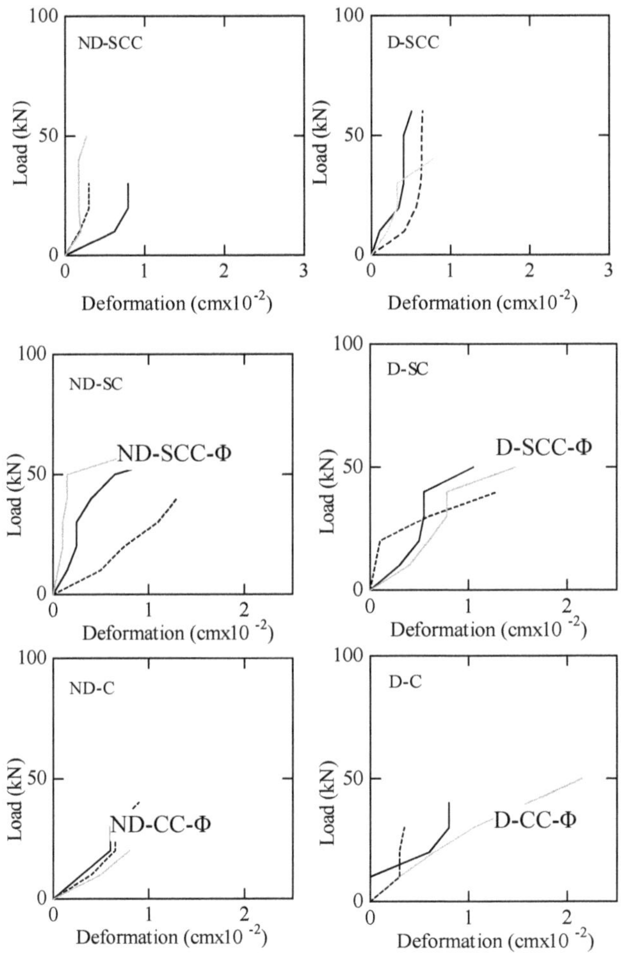

Figure 4: Load-deformation relation of wall specimens strengthened by self compacting concrete and standard concrete.

loading is observed in ND-CCΦ brick wall samples, i.e., undamaged samples strengthened by standard concrete with steel wire mesh reinforcement. On the other hand, the performance of the reinforced walls is quite acceptable. It is likely that desirable results can be obtained when strengthening is provided on one side of the walls of the masonry structures using self-compacting concrete with reinforcement. Taking into consideration the fact that the adjoining building will only allow for strengthening on the inner sides, it is clear that this will also provide considerable improvement in the load bearing capacities. Even without reinforcement, the strengthening with self compacting concrete provides significant improvements in strength of masonry walls.

Table 1: General experimental results.

<table>
<tr><th colspan="3" rowspan="2"></th><th rowspan="2">Specimen</th><th colspan="2">Max. strength</th></tr>
<tr><th>Average (MPa)</th><th>COV (%)</th></tr>
<tr><td rowspan="4">SELF COMPACTING CONCRETE</td><td rowspan="2">WITHOUT REINFORCEMENT</td><td>Damaged</td><td>D-SCC</td><td>4.01</td><td>25.1</td></tr>
<tr><td>Non Damaged</td><td>ND-SCC</td><td>3.62</td><td>21.27</td></tr>
<tr><td rowspan="2">WITH REINFORCEMENT</td><td>Damaged</td><td>D-SCCΦ</td><td>4.19</td><td>17.48</td></tr>
<tr><td>Non Damaged</td><td>ND-SCCΦ</td><td>4.91</td><td>12.11</td></tr>
<tr><td rowspan="2">STANDARD CONCRETE</td><td rowspan="2">WITH REINFORCEMENT</td><td>Damaged</td><td>D-CCΦ</td><td>3.74</td><td>18.04</td></tr>
<tr><td>Non Damaged</td><td>ND-CCΦ</td><td>2.94</td><td>21.84</td></tr>
<tr><td colspan="3">UN-REINFORCED MASONRY WALL</td><td>URM</td><td>1.53</td><td>15.9</td></tr>
<tr><td colspan="3">DAMAGED-EPOXY REPAIRED URM</td><td>D-ER</td><td>2.98</td><td>20.5</td></tr>
</table>

Experiments have been performed on 18 brick wall samples in 6 different series strengthened by standard concrete and self-compacting concrete. Fig. 5 illustrates the average strength of the brick wall samples series.

Compared to those strengthened with standard concrete, the behaviour of non-damaged brick wall samples strengthened with self compacting concrete change to a great extent. The increases in failure loading capacity stand between 100% and 170% when strengthened with conventional concrete and with self compacting concrete, respectively.

No significant improvement has been achieved with the pre-damaged brick wall samples close to their failure points and repaired with epoxy resin, even if they are strengthened with self-compacting concrete. In some cases, brick wall samples strengthened with the conventional concrete, however, exhibit slight improvement. The behaviour of the brick wall samples is closely related to the status, character, and number of the cracks formed during pre-loading of the samples and the quantity of the epoxy resin used in repairing the cracks formed. The brick wall samples suffer damage at their weakest portion during the pre-loading process carried out to induce damage. These cracks are generally repaired along the compression diagonal and that provides significant increase in the strengths of these portions compared to the other portions. It is therefore highly likely that this, rather than the type of the concrete used in strengthening is the main factor defining both the failure and deformation behaviour of the samples.

Another point that deserves emphasis here is that the sizes of the brick wall samples are small and therefore, the repaired portion of the wall covers the majority of the wall surface area. The improvement obtained following strengthening with the self-compacting concrete is evident when observing the behaviour of non-damaged brick wall samples. Self-compacting concrete has advantages over conventional concrete with regard to better encasing of the reinforcement, providing a better bond between the reinforcement layer and the wall surface, self-compacting and self-levelling and featuring a nonporous wall surface area. It is evident that the role of the repair with epoxy resin will decrease as the walls become bigger or reach actual size.

The improvement obtained by strengthening with the self-compacting concrete is obvious in the behaviour of non-damaged wall samples. This is due to the better encasement of reinforcement by the self-compacting concrete, the stronger bond between self-compacting concrete and the wall, the self levelling advantage and the nonporous structure of the material. Higher deformation caused by self-compacting concrete can be explained by the fact that self compacting concrete exhibits more ductile behaviour compared to standard concrete.

4 Case study

In Kayseri-city, a site conservation area with seven stone houses built around 1900 was restored by the Kayseri Metropolitan Municipality. Some of these are mansions and consist of 3 floors including a basement, a ground floor and a first floor. All buildings have a flat roof and built contiguously (fig. 5).

The historical buildings under preservation consisted of tokana/traditional kitchen, hamam/bath, köşk/kiosk, haremlik/woman section of the house, selamlık/men section of the house and sofa/main hole and represent many historical signs through their banisters, wood carvings and masonry ornaments.

For the repairing of cracks on masonry walls at the basement and upper floors, epoxy resin injection was used partially and for the inner walls CFRP was applied. CFRP application was accepted by the authority in charge for the inner walls and the inner sides of external walls as they were later covered with timber lining.

Kuyumcuoglu Mansion, which is one of the restored stone houses in question, had out-of-plane deformations on the front face of ground floor walls caused by the inner side section loss. The deformation does not seem to affect the

Figure 5: Front wall external view.

appearance of the building; however repair was unavoidable as the safety of the structure was threatened.

Section loss in question reduced the thickness of the front walls from the original 25 cm to 10–12 cm. Application of CFRP was not a suitable solution because of the extreme section loss. What should have been done here was to keep the original front face and thicken the section using similar stones in the inner side. However, as the original part of the wall was insufficient for bearing loads, it was decided to use self-compacting concrete of 10 cm thickness on wire mesh reinforcement and cover the inner surface with a thin layer of the original stones which makes the total thickness 25 cm again. Wire mesh reinforcement was fixed on the inner surface by using steel connectors and epoxy resin. During the application of self-compacting concrete, the front face of the wall was supported to prevent any collapse.

5 Conclusions and discussion

The URM walls usually collapse upon the formation of cracks along the loading axis or else, in line with the joints or in line with both the orientations of joints and the bricks (fig. 6).

In some cases specimens break into pieces. It is established that strengthened brick wall samples were not broken into pieces. Following strengthening, URM wall samples feature significant increase in their load-bearing capacities. On the other hand, the deformations of the brick wall samples decrease to a great extent.

When strengthening is applied to the brick wall samples, the load-bearing capacities tend to become dependent on numerous parameters. It is therefore difficult to state the exact improvement ratios. For the purpose of approximation, it can be concluded that the load-bearing capacities of the URM wall samples strengthened on one side are increased by 170% when strengthened with self compacting concrete.

Strengthening of pre-damaged brick wall samples by self-compacting concrete provides significant change in their behaviour, as compared to the standard concrete. Even without reinforcement, the strengthening with self compacting concrete provides significant improvements in strength of masonry walls. This positive change is not, however, observed in some pre-damaged samples. The behaviour of the brick masonry wall samples is closely related to status, character, and number of the cracks formed during pre-loading of the samples and the quantity of the epoxy resin used in repair of the cracks formed. Repair of these cracks are generally performed along the compression diagonal however, this ends up with significant increases in their strengths of these portions compared to the other portion of the samples. It is likely that the failure or deformation behaviour of the samples is determined by that, rather than the type of the concrete used in strengthening.

Figure 6: Failure mode of a URM and strengthened specimen.

Another point to be emphasized here is that the sizes of the brick masonry wall samples are small and therefore, the repaired portion of the wall covers the majority of the wall surface area. It is evident that the role of the repair with epoxy resin will be reduced for bigger or true size walls. The improvement obtained following strengthening with the self-compacting concrete is evident when observing the behaviours of non-damaged brick masonry wall samples.

Where the samples are strengthened by the standard concrete, no significant difference is observed between the load-bearing capacities of pre-damaged and non-damaged brick wall samples. In fact, the samples strengthened without pre-damaging would be expected to perform better, however it is likely that the repair with epoxy resin contributes to the damaged samples' strengths in excess of the non-damaged ones.

The fact that the samples' surfaces are repaired with epoxy resin reduces the concrete adherence to the surfaces; this gives rise to such peculiar behaviour.

The inclusion of the steel wire mesh reinforcement in the concrete enhances this effect. This is mainly because the loss of adhesion may give rise to premature separation that prevents the occurrence of the expected behaviour.

The impermeability of the self-compacting concrete is twice as much as compared with the standard concrete. In strengthening the masonry structures, the service life will be improved, in addition to the increase of load-bearing capacities.

Furthermore, it is possible to obtain superior surface smoothness in strengthening as compared with the standard concrete, thanks to the self-compacting concrete's self-placing, self-levelling and nonporous strengthening properties. This also eliminates the need for the plaster, thus reducing the time and labour for strengthening. In the strengthening process, the quality attributes are maintained during handling and settling of the concrete, which might be preferable. Furthermore, it is established that the adherence obtained between the layer and the brick wall is better in case of the self-compacting concrete as compared to the standard concrete.

On the other hand, the adherence might adversely be affected when the epoxy resin is used in repairing pre-damaged walls. This might, however, be just the opposite as well. It is therefore necessary to roughen the surfaces repaired with epoxy resin by making use of special process to increase the adherence with the concrete.

Acknowledgements

The study presented in this article was sponsored by Lafarge-Concrete Turkey. The authors acknowledge the financial support provided and the Self Compacting Concrete (Agilia) supplied.

References

[1] Ediz, I., An experimental study on retrofitting of structural brick masonry walls using steel wire mesh reinforcement and self compacting concrete, M.Sc. Thesis, Institute of Science and Technology, Istanbul Technical University, 2006.

[2] Shing, P.B., Noland, J.L., Klamerus, E. & Spaeh, H., Inelastic behavior of concrete masonry shear walls. *Journal of Structural Engineering*, **115(9)**, pp. 2204-2225, 1989.

[3] Drysdale, R. G. & Khattab, M.M., In-plane behavior of grouted concrete masonry under biaxial tension-compression. *ACI Structural Journal*, **92(6)**, pp. 653-664, 1995.

[4] Agarwal, P. & Thakkar, S.K., A comparative study of strengthening and retrofitting measures for unreinforced brick masonry model under cyclic testing. *Journal of Earthquake Engineering*, **8(6)**, pp. 839-863, 2004.

[5] American Society for Testing and Materials (ASTM). Standard test method for diagonal tension (shear) in masonry assemblages. Annual book of ASTM standards, E 519-81, Vol. 04.05,1981.

Reconstruction of the Rector's Palace atrium in Dubrovnik: the key role of column connections

D. Lazarević, J. Atalić & K. Fresl
Faculty of Civil Engineering, University of Zagreb, Croatia

Abstract

The Rector's Palace is one of the most valuable heritage buildings in the historic Old Town of Dubrovnik and the world famous setting for concerts and exhibitions. The Palace Atrium is surrounded by a two-storey structure comprising quadripartite masonry vaults supported by masonry walls and stone columns. Historically, this structure has suffered numerous collapses and damage and restorations have frequently been made. Despite this, many distinct fractures are concentrated around the traditionally made column connections with capitals and bases. Carefully smoothed stone contact areas are joined with iron dowel, which is centrally placed in a slightly larger hole, with a lead infill. These contacts do not have sufficient rotational stiffness and the ability to transmit eccentric loads. Therefore, high edge stresses and fractures at connections are unavoidable. With time, the small rotational stiffness of joints becomes even smaller and the behaviour is approaching one of the hinge joint. This effect may endanger the global stability of this structure, as only three aligned hinges are sufficient to render the system unstable, i.e. the system is "almost" statically determinate and has only a small reserve in bearing capacity. Various retrofit strategies for column connections can cause considerably different stress distribution and by ignoring or underrating these effects, the safety factor can be seriously overestimated. This paper confirms the above assertions using a range of computational models: from simple frame and shell models to complex 3D solid models. This work also points out the significance of connections in historical buildings and attempts to investigate possible retrofit concepts. Numerous retrofit solutions were considered and results show a significant influence of this small detail of the bearing system on the global behaviour and stability of structure.

Keywords: historic stone masonry, columns connections, dowel, contact stress.

1 Introduction

The Rector's Palace is one of the most outstanding monuments of Gothic and Renaissance architecture, not only in Dubrovnik, but also on the entire Adriatic coast. It is placed on the list of protected cultural monuments, established by the UNESCO organization. The written history of the Palace dates from the 13th century and it is fairly well known that the present day appearance is a result of numerous events that occurred throughout several centuries [1]. Today, the Palace is a tourist and cultural centre, the location for various exhibitions and a world famous place for chamber music concerts. However, numerous fractures, damaged columns, stirrups, insets and broken parts seriously decrease the global impression of this monument. To illustrate the persistence of the problem, two characteristic quotations, found in the archives [2], could be mentioned: "*... during the summer festival 1968 part of the capital fell down ...*" and "*... this is part of the capital from the left column near the clock, which was replaced in 1952*". Some historical data and numerous vestiges of previous reconstructions show that a serious problem needs to be addressed. In this paper, the fundamental causes of permanent damage of column elements in the atrium of the Rector's Palace are considered.

1.1 Description of the atrium

The Rector's Palace (fig. 1(a)) is a traditional structure placed inside the historical part of Dubrovnik, called the Old Town. The massive stone walls of the Palace surround the inner opened structure, known as the atrium (fig. 1(b)) where the origins of the problem are located. It consists of two storeys; each defined by the quadripartite masonry vaults (fig. 1(c)) and filled with a weak material up to the floor level. The back side of the vaults is supported by masonry walls, while the front side lies on the stone columns, which mostly constitute the inner façade (fig. 2(a)). The first floor columns are coupled and have smaller diameter (d = 23–25 cm) than those on the ground floor (d = 50 cm). Between the columns, which carry the floor vaults, there are also columns that carry only decorative front arches. On the northern front, instead of the row of columns, ground floor vaults lie on a long masonry wall, along which the stone stairway leads to the floor level (fig. 2(b)). At the corners, the floor vaults are supported by massive brick columns marked in fig. 2(c).

Figure 1: Rector's Palace: (a) front view, (b) atrium (ground floor), (c) atrium (first floor).

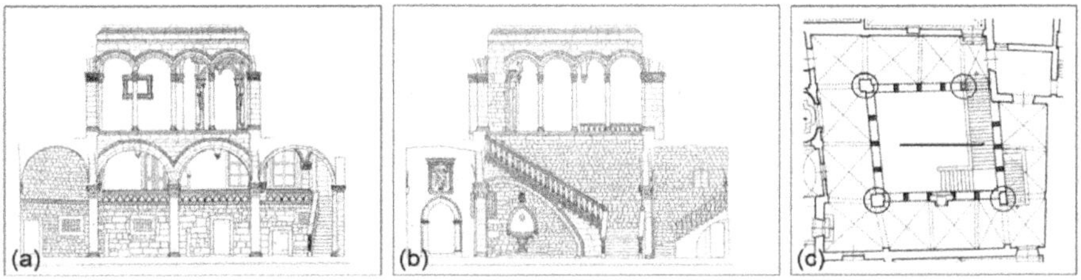

Figure 2: Atrium: (a) south front, (b) north front, (c) plan of the floor.

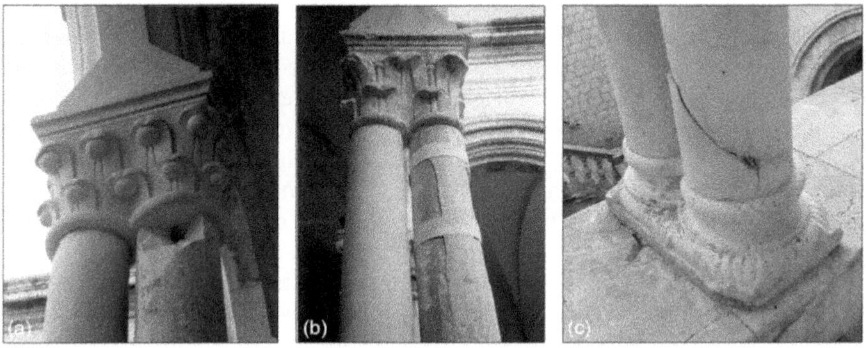

Figure 3: Fractures on column connections: (a) and (b) with capital, (c) with base.

Two additional bearing elements should be mentioned: braces below the floor arches and the reinforced concrete grillage placed in the infill of the terrace floor, both parts of previous reconstructions.

1.2 Current condition of the structural system

Centuries will cause deterioration in any structure. From historical reports and discussions with colleagues from various fields, who have worked on the Rector's Palace during the past four decades, it became apparent that the structure has suffered numerous damage throughout history. Prompted by the characteristic fractures on the bases and capitals of many columns (fig. 3), a working hypothesis was made that the reasons of past collapses partly lie with the weak joints between the column elements. Indeed, many restorations were noted by chronicles and today numerous signs of these actions can be observed: different types of stones, various styles of column shapes and capitals, stirrups around columns, stone insets and gaps filled with led or wood. Wrong assertions about large compressive force in columns as the main reason of their collapse compelled old masters to remove a heavy stone fence from the terrace. Almost forty years ago, structural engineer V. Steinman wrote in his report [2]: "*Damages of particular parts of columns seem to represent a continuous phenomenon which required unusually large number of interventions*".

Today, it is obvious that the braces are not correctly placed, activated and anchored, a few capitals are eccentrically loaded, and the beneficial action of any axial load due to weight of the fence is reduced. If the event of a gunpowder explosion in 1463, serious earthquakes in 1520, 1667, and 1979, many reconstructions (major in 1981), and presumably lots of unknown interventions are added, it is obvious that the underlying bearing system is seriously disrupted.

2 Problem statement and hypothesis about structural behaviour

The logical hypothesis about large relative vertical displacements between the columns, as the main reason for damage, was not confirmed by observations, geotechnical and surveying measurements. However, this statement is only conditionally discarded because of the short measurement time scale and the possibly slow (fig. 4(a)) or discontinuous (fig. 4(b)) ground displacements, where the latter appeared immediately after a seismic event. These scenarios offer an explanation why precise instruments did not detect any movement over the period of a year or two.

Another hypothesis about the large uniform compressive stress in columns was also discarded, based on both experimental as well as numerical findings. Therefore, the solution to the problem was pursued further. Although there are "standard" reasons for deterioration in historical constructions, which can always be used as the "reasons for damage" (dowels corrosion, for example), preliminary results pointed to a specific problem of column connections.

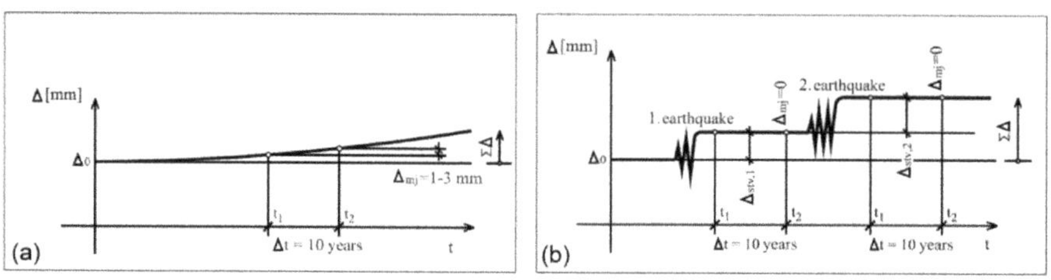

Figure 4: Possible long time displacement of support: (a) slow and continuous, (b) immediate and discontinuous.

2.1 Connection properties

Detailed analysis of specific fractures around the column connections revealed some bad features of contact between the column and the base as well as the column and the capital. Carefully smoothed stone contact areas are joined with an iron dowel which is centrically placed in a slightly larger hole and finally filled with lead (fig. 5(a)). Old masters must have thought that the columns were uniformly compressed, so such precise finishing of contact areas was very well suited for transmission of such uniform vertical stresses.

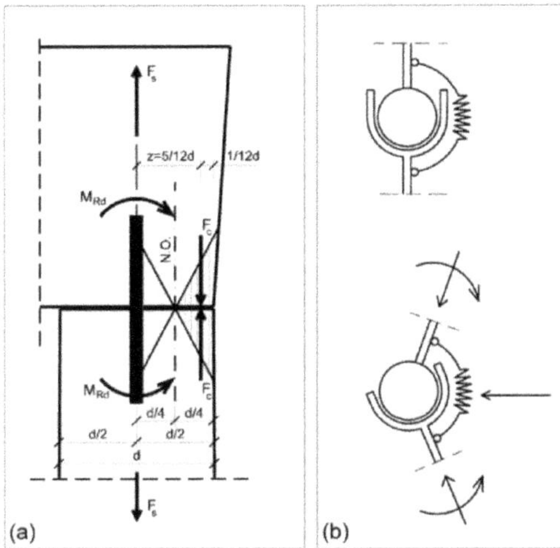

Figure 5: Connection contact model: (a) rotational capacity, (b) undeformed and deformed state.

Unfortunately, it will be shown in the next section that they were completely wrong. Furthermore, the dowel can transmit only a very small tensile force due to insufficient dowel anchoring (about 10 cm) and the lack of adhesiveness amplified by the influence of weak mechanical properties of lead, especially its tendency to creep. Therefore, the dowel can safely maintain only compression and transverse force (with a help of friction between stones at contact). It is obvious that such a connection has a very small rotational stiffness and therefore the contact behaves almost like a hinge (more precisely, like a hinge with an added rotational spring with small stiffness). So, contact behaviour can be primarily described like a ball-in-a-cup (fig. 5(b)). Note, even intuitively, that such a contact is very sensitive to any horizontal disturbance.

2.2 Eccentrically loaded joints

Unfortunately, connections are mostly loaded eccentrically (note the position of the thrust line in fig. 8(b)). Even small horizontal displacement of connection causes relative rotation between column elements and results in high local stresses at the edge of a column, because the joint has only very limited ability to compensate rotations (fig. 6). When stress concentration exceeds the compression strength of stone, the deviation of compression trajectories generates an unwanted tensile force which easily exceeds the small tensile strength limit of stone and causes a column to split (fig. 7a). This force is effectively covered with stirrups (in reinforced concrete) or with steel rings (as it appears here), tightened around almost crumbled columns. Eventually, this leads to fractures near contact surfaces, splitting of columns and capitals and thereby a complete loss or a very significantly reduced bending stiffness.

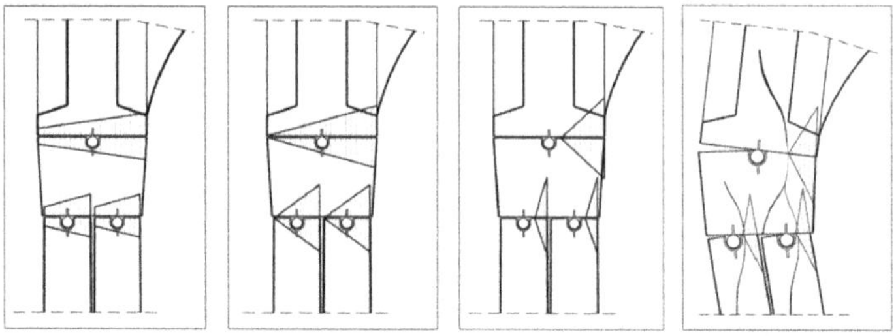

Figure 6: Progress of contact stresses with an increase of lateral load.

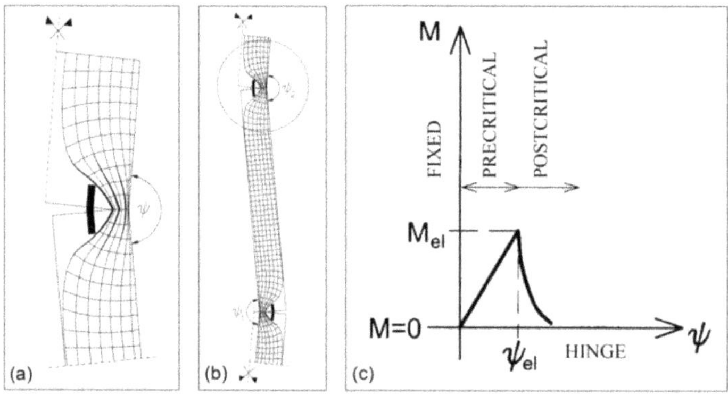

Figure 7: Contact behaviour: (a) position of cracks, (b) stress trajectories in column (c) approximate M – ψ diagram of contact.

The connection is still capable of transmitting compressive and shear forces and therefore can be idealised as a pure hinge. It should be noted that this problem is very localised as outside a range of approximately $3d$ from the joint, the trajectories become parallel again (fig. 7(b)). Consequently, the body of the column experiences only moderate uniform compression stress and a large safety factor is maintained.

2.3 Lack of redundancy

However, if the connections are idealised as hinges (fig. 8(a)), the eccentrically loaded bearing system (fig. 8(b)) can easily turn into a mechanism. Note that only three hinges on the characteristic vertical line are sufficient to render the system unstable (fig. 8(c)). Eventually, most of the connections are seriously cracked (on the descending part of $M - \psi$ curve; see fig. 7(c)) and the global bearing capacity of the atrium is almost exhausted (fig. 8(d)). The structure does not have bearing capacity reserves and the overstressed joints cannot be effectively unloaded. High contact stresses cannot be redistributed to neighbouring bearing elements due to the small degree of indeterminacy. It is interesting to note that the computed crack positions (fig. 8(e)) are in agreement with those found on the real structure.

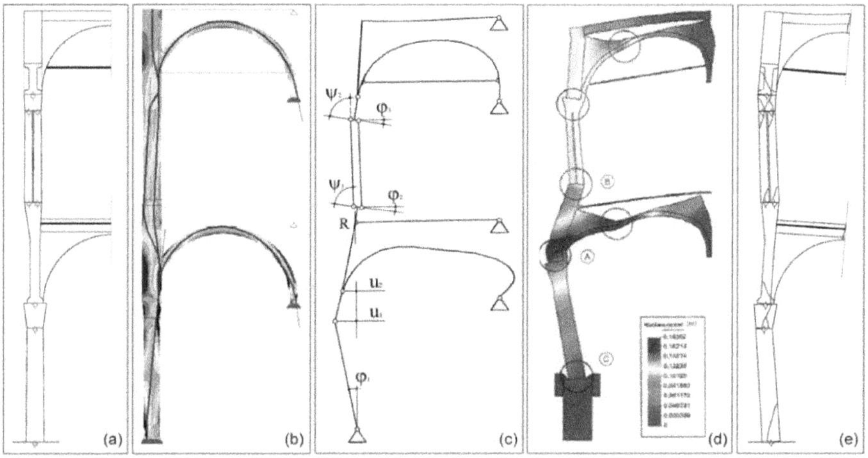

Figure 8: Behaviour of characteristic section: (a) position of hinges, (b) thrust line, (c) frame model, (d) stresses in section, (e) calculated cracks.

It can be concluded that the system equilibrium is maintained in the fractured state, the stability of which depends on the weakest material property of the stone – the small tensile strength which ensures only a weak fixity of joints. Without it, the structural system would have sufficient number of hinges to become a mechanism.

2.4 Additional difficulties

The state of equilibrium, which is maintained only by the rotational stiffness of cracked stone is very sensitive to any load changes. Even everyday temperature gradient causes both increase as well as decrease of the compression zone and the corresponding changes of splitting forces directly influence opening and closing of cracks. Therefore there is constant activity within these connections and the associated propagation of fractures (fig. 9). If a large vertical load were to exist, it would compress the joints and it would centre the thrust lines, thereby minimizing the problem of the local overstress. Unfortunately, the measured values show a low level of compressive stress in the columns. Axial forces in the columns were additionally reduced by the removal of massive stone balustrade from the atrium terrace in 1952. The trouble is that such weakly compressed columns are also very slender, having small core sections and therefore are highly sensitive to any load eccentricity.

Additional stresses due to the pressure of corroded dowels are also present. It is well known that the corrosion causes an increase in the dowel volume and hence local pressure and splitting forces are significantly magnified. Braces which were added below the terrace vaults are much more efficient in preventing the horizontal displacements, rather than the rotation angles.

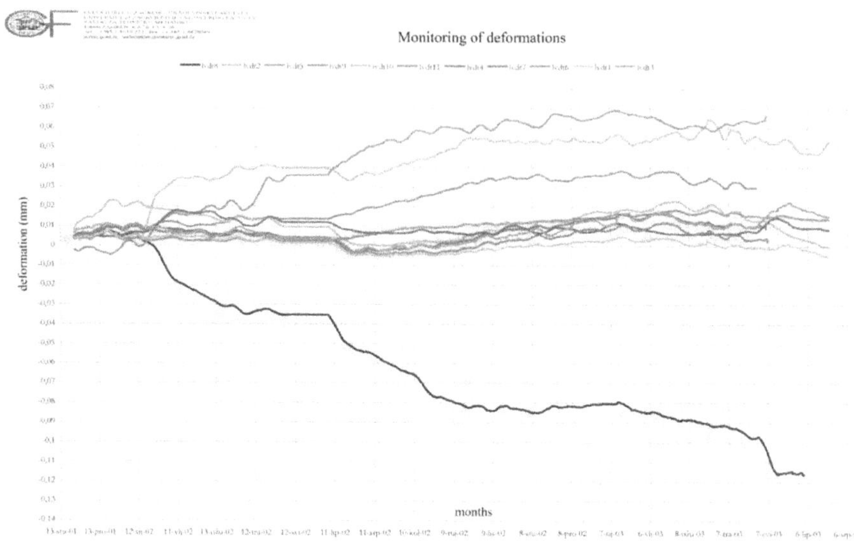

Figure 9: Monitoring of active cracks as measured during period of 22 months.

Their bending stiffness is negligible to prevent the rotation of the anchored place. It should be emphasized that these displacements and rotations are very small. Computational simulation indicates that the first crack opens with the horizontal displacement of the contact of less than one millimetre.

Further concern arises from the fact that Dubrovnik is placed in a region of high seismicity, with an expected peak ground acceleration of around 0.4g. In 1981 the Rector's Palace was reinforced to sustain moderate earthquake loads, but since then, new regulations require consideration of significantly increased loads. Massive walls around the atrium should be able to sustain most earthquake forces but will also displace horizontally. As the walls are connected with vaults, these displacements will be common and will affect the atrium. Therefore, rotations of the column elements will invariably additionally increase.

It should be mentioned that the role of fracture mechanics is significant here. Even an earthquake of lower intensity produces enough energy to exceed a small fracture energy of a brittle material like stone. Energy needed for fracture propagation is proportional to the volume of the damaged area. So, fractures can easily propagate through slender and fragile elements like the tiny columns in the atrium.

3 Numerical modelling strategies

Various computational models have been constructed to support the above statements and to test the effects of rotational stiffness on the stress distribution and particularly on the local and global stability of the structure.

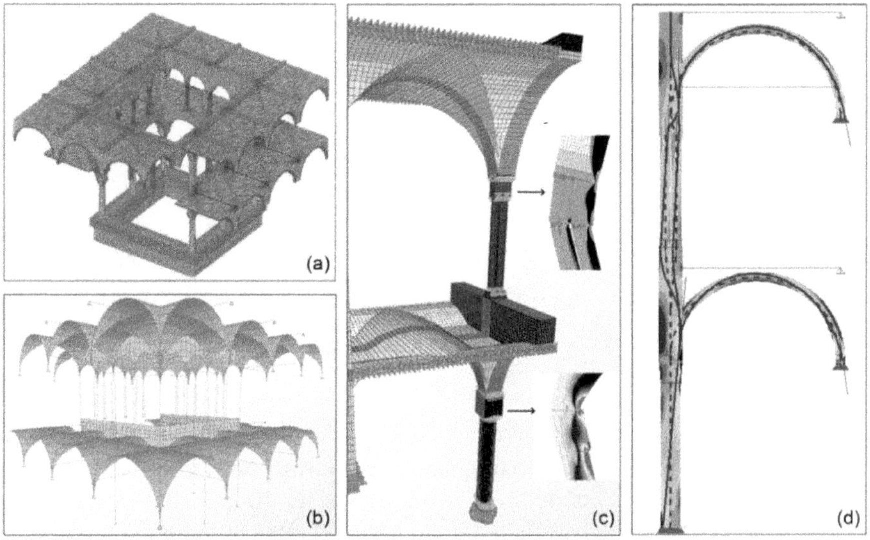

Figure 10: Computational models: (a) with brick elements, (b) with frame and shell elements, (c) contact stresses, (d) lines of thrust.

3.1 Model of complete structure

A computational model of the complete structure, with tetrahedral finite elements, was analysed using FEAP 7.4 [3] combined with GiD 6.1.2a [4] for pre and post processing purposes (fig. 10(a)). Several novel subroutines were implemented into the basic linear elastic FEAP model to better describe the connection behaviour. The program code was extended to exclude overstressed finite elements and to search for the minimal energy in a given direction (line search) using the bisection method to improve the convergence of the incremental Newton–Raphson technique. For the purpose of numerical integration over the tetrahedron (with ten nodes and three degrees of freedom per node), four Gauss points were used.

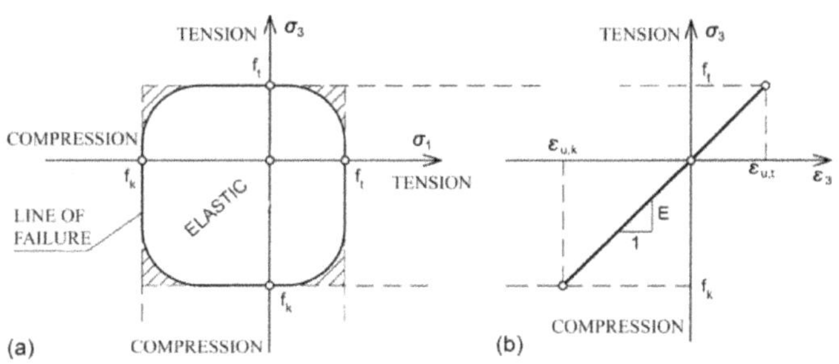

Figure 11: Fracture criterion: (a) failure surface, (b) one dimensional σ-ε curve.

The failure criterion of the material was defined by the modified theory of normal stresses, which was originally developed by Galileo and Rankine. This simple theory predicts the failure of an initially elastic material, with no ductility and no post critical reserves. The fracture criterion is defined by the expression:

$$f(\sigma_1, \sigma_3) = \frac{(\sigma_1 - \sigma_a)^n}{\sigma_s} + \frac{(\sigma_3 - \sigma_a)^n}{\sigma_s}.$$

where σ_1 and σ_3 are the highest and the smallest principal stress at the Gauss point of an element. Here, a positive value implies tensile, and a negative value compressive stress (fig. 11). The values σ_a and σ_s are defined by:

$$\sigma_a = \frac{f_t - f_k}{2}, \qquad \sigma_s = \frac{f_t + f_k}{2}.$$

Based on this criterion, materially nonlinear analysis with load increments of 1% of the total load was conducted. Within each step, the full Newton–Raphson method was applied. The stability of the analysis was monitored by the line search method, using the bisection for computing a zero point and an under-relaxation factor equal to 0.8 was adopted for additional stability purposes. The convergence criterion was based on an energy norm. Convergence was deemed to have been achieved and the next step was started if the current energy was lower than 10^{-5} of the energy at the start of the step. Using this algorithm, stress concentrations and the exclusion of overstressed elements were easily detected. Degradation of stiffness was monitored by plotting the relation between the first eigenvalue of tangent global stiffness matrix and the safety factor for estimated dead load (fig. 12). After three substantial step changes in the eigenvalues, here labelled as A, B and C (three hinges were developed), the global stiffness was so small that the model became practically unstable [5].

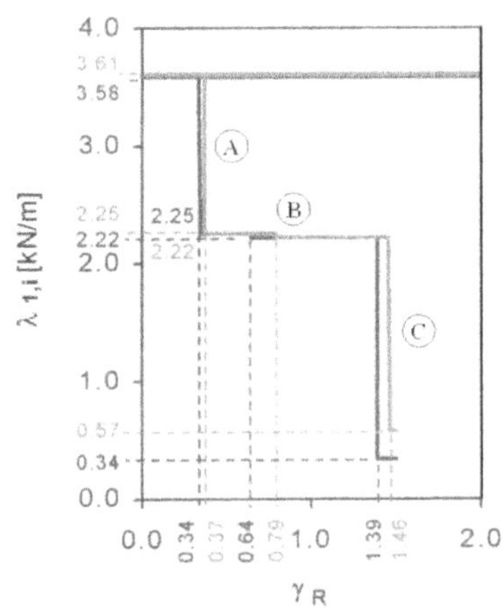

Figure 12: Decrease of the first eigenvalue of the global tangent stiffness matrix.

3.2 Detailed model of the characteristic cross section

Detailed modelling of the column connections was preformed with SAP2000 [6] using brick elements with incompatible modes and 'links' (gaps) elements, which are able to exclude tension stresses (fig. 10(c)). Minor fractures in the vault bearing system were ignored and links were used only at column connections including properties of lead and the dowel. Even this simple approach was sufficient to obtain contact rotations and overstressing at the edges. Behaviour and positions of overstressed elements matched fairly well observed fractures.

Further computational models of various complexities were explored to assess the suitability of simplified models for general usage. For example, the same gap elements were used for the contact behaviour on a simple planar model of the characteristic section (fig. 10(d)). In this model, the gap opening and its influence on the stress distribution and the lines of thrust (continuous and dashed lines) are easily followed. Also, the effect of contact properties on the shape of the thrust line can be seen, if the position of centred (dashed) and inclined (continuous) line are compared. It is easy to see that the dashed thrust streamlines are in agreement with high compressive stresses. Completely different behaviour can be expected if the line of thrust is centrally placed. In such a case, the best of the stone properties – its great compressive strength - is beneficial.

Regardless of the level of sophistication of the chosen computational setting, the line of thrust should still be used as a fundamental technique for preliminary analysis of historical constructions. This simple and approximate approach always shows main structural weaknesses that cannot be ignored.

3.3 Simplified model

A mixed frame and shell model of the complete structure (fig. 10(b)) was finally built to study the influence of different rotational stiffnesses obtained from detailed models. The columns were modelled with frame elements and rotational stiffness between them was included in the form of a rotational spring. The effects of previously known or recorded reconstructions were included approximately by the staged construction technique to correctly activate subsequently inserted elements. All analyses accounted for geometrical nonlinearity through the *P*-delta effect. As the level of the rotational stiffness changed, the model behaviour and the results were influenced significantly. The global model is highly sensitive to the changes on local contact properties. This approach provided insight into the stability problem and the evaluation of the safety factor. Another important benefit was the evaluation of maximal angles of rotations at joints, which identified the size of required dilatation needed for the connections repair [7].

Various models confirmed the importance and the need for the proper assessment of the column connections. The level of rotational stiffness was crucial for the understanding of load carrying mechanisms and for finding a correct distribution of stresses and assessment of the global stability of structure.

It is hard to present a unique rule for the analysis of column contacts because there are various types of columns, joints and structural systems. However, the importance of details can be emphasized by the fact that the safety factor for this structure dropped from 10 for standard continuous elastic model to 1.4 for models based on the assumptions presented.

4 Possible retrofit strategies

Analyses results confirmed our assertions about problematic column connections and the need to act in order to preserve both the local and global stability of the structure. The previous history of the Rector's Palace is ripe with inadequate interventions and one of the goals must be to extend the time without new fractures of columns as much as possible. Some form of repair (or even replacement) of critically damaged columns and capitals is clearly inevitable.

In order to act on the side of caution and to avoid potentially suggesting yet another problematic intervention (recognising also that there are a number of other similar cases in Croatia) all the above mentioned assumptions were confirmed with laboratory tests [8] in the context of an associated comprehensive research project. Tests were carried out on nearly full scale stone column samples, supplied from traditional quarries (fig. 13(a)). Experiments included both concentric and eccentric loading, reflecting the actual load distribution in the structure, as well as conditions associated with the most important stages of the proposed interventions.

Key intervention will include applying a thin layer of lead (about 5 mm) placed between the two connected areas (fig. 13(b)). A material such as lead will enable small rotations and will shield connected areas from large stress concentrations. Lead is already used to fill dowel holes and the additional horizontal thin layer appears as an appropriate and almost invisible intervention. In order to reduce load eccentricity in columns, which primarily exists due to thrust of first floor vaults, new braces are envisaged. These will be slightly prestressed (30 kN) and properly anchored. This visible (and frequently unpopular) intervention is needed if the thrust line is to be centred, which will in turn minimize the consequences of serious earthquakes. Furthermore, existing braces and anchors from earlier interventions at the second floor are overstressed and must be replaced in any case. This issue is absolutely vital, as without them, the thrust line lies outside the column section and the structure is unable to sustain even a dead load.

All potential reconstruction phases were also checked using a nonlinear staged construction model, with element inclusion/exclusion when required as the analysis progresses in time. These elements were carefully placed in such a way that loads on structural elements in all intervention phases were on the conservative side (fig. 13(c)). With these interventions, the safety factor of the structure is increased by approximately 50% and it is hoped that any future interventions on columns will be significantly delayed.

Continuous, long term monitoring is imperative because the presented analyses are based on a number of unknown and/or assumed data. As, for

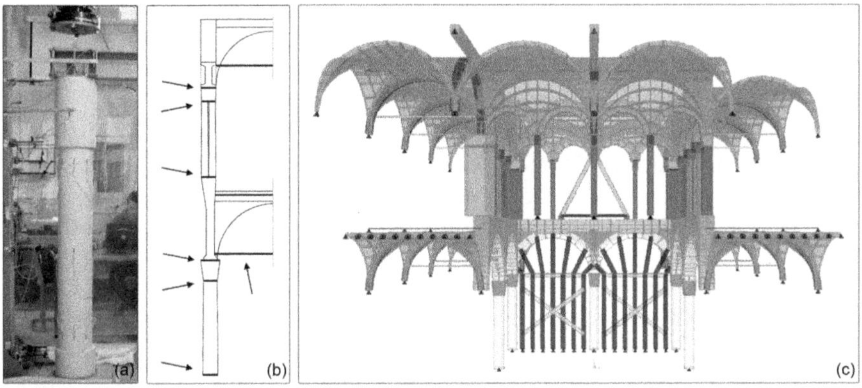

Figure 13: Reconstruction: (a) laboratory test (b) locations of intervention, (c) numerical model.

obvious reasons, destructive testing methods are not appropriate for historic buildings; insufficient data usually exists to characterise constitutive model parameters. Therefore, non-destructive experimental strategies must also be considered and/or further refined.

5 Conclusion

The importance of correct analysis of stone column connections is emphasized, along with the complexities and difficulties associated with such a process. Slender columns are usually identified as the weakest link in historical constructions. If the hypotheses about the uniform stress distribution cannot be proven for a specific situation, it is vital to fully appreciate the behaviour of the adopted contact details. It was shown that, for the analysed structure of the Rector's Palace Atrium in Dubrovnik, the joints lack sufficient rotational stiffness and the eccentricity effects due to dead load (inclined thrust lines) cause fractures at places of contact. It is concluded that ignoring or not fully appreciating the real nature of connections leads to significantly different stress distributions and orientations of thrust lines which may grossly overestimate the safety factor. Moreover, the propagation of fractures near the column connections can easily endanger the global stability of structure.

References

[1] Horvat-Levaj, K. & Seferović, R., Baroque restoration of the Rector's Palace in Dubrovnik. *Radovi Instituta za povijest umjetnosti*, **27**, pp. 163-168, 2003.

[2] Steinman, V., *Research Work on Rector's Palace in Dubrovnik with Basic Recommendations for Reconstruction*, (in Croatian) Book II, Civil Engineering Institute of Croatia: Zagreb, pp. 50-83, 1974.

[3] Taylor, R.L., *FEAP – A Finite Element Analysis Program, Version 7.3, Programmer Manual*, Department of Civil and Environmental Engineering, University of California: Berkley, pp. 20-48, 2001.

[4] GiD, *The universal, adaptive and user friendly pre and post processing system for computer analysis in science and engineering,* Reference Manual Version 6, CIMNE: Barcelona, pp. 127-141, 2000.

[5] Lazarević, D., Dvornik, J. & Fresl, K., Analysis of damages to the Rector's Palace atrium in Dubrovnik, *Građevinar*, **56(10)**, pp. 601-612, 2004.

[6] SAP2000, *Intergrated Software for Structural Analysis and Design,* Analysis reference manual, CSI: Berkeley, pp. 189-224, 2002.

[7] Atalić, J., Lazarević, D. & Fresl, K., Influence of rotational stiffness between column elements on global stability of historical constructions. *Proc. of the 8th European Conference on Research for Protection, Conservation and Enhancement of Cultural Heritage: book of abstracts*, eds. J. Kolar, & M. Strlić, Narodna in univerzitetna knjižnica: Ljubljana, pp. 60-62, 2008.

[8] Lazarević, D., Atalić J., Krolo J., Uroš M. & Šavor M, Experimental and numerical analysis of traditional column connections with the possible retrofit concept. *Advanced Materials Research*, **133-134**, pp. 479-484, 2010.

[9] Lazarević, D., Dvornik, J., Fresl, K. & Rak, M., Numerical analysis of damages of the Rector's Palace atrium in Dubrovnik. *Proc. of the Int. Conference Heritage Protection - Construction Aspects*, eds. J. Radić, V. Rajčić & R. Žarnić, SECON HDGK: Dubrovnik, pp. 109-116, 2006.

[10] Palmisano, F., Vitone, A. & Vitone, C., Load path method in the interpretation of masonry vault behaviour. *Structural Studies, Repairs and Maintenance of Heritage Architecture IX,* eds. C.A. Brebbia & A. Torpiano, WIT Press: Southampton, pp. 155-168, 2005.

[11] Meli, R. & Sanchez-Ramirez, A.R., Structural aspects of the rehabilitation of the Mexico City Cathedral. *Int. Sem. Structural Analysis of Historical Constructions. Possibilities of numerical and experimental techniques*, eds. P. Roca, J.L. Gonzales, A.R. Mari & E. Onate, CIMNE: Barcelona, pp. 123-140, 1997.

[12] Theodossopoulos, D., Structural design of high gothic vaulting systems in England, *International Journal of Architectural Heritage*, **2(1)**, pp. 1-24 2008.

[13] Macchi, G., Problems related to the original conception – The case of Pavia Cathedral. *Int. Sem. Structural analysis of historical constructions II. Possibilities of numerical and experimental techniques*, eds. P. Roca, J.L. Gonzales, E.Ońate & P. Lourenço, CIMNE: Barcelona, pp. 123-140, 1998.

[14] Binda, M.L., Saisi, A., Messina, S. & Tringali, S., Mechanical damage due long term behaviour of multiple leaf pillars in Sicilian Churches. *Proceedings of the 3rd International Seminar Historical Constructions 2001. Possibilities of numerical and experimental techniques,*

eds. P. Roca & P. Lourenço, University of Minho: Guimaraes, pp. 707-718, 2001.

[15] Gubane, A., & Giuriania, E., Structural restoration of vaults with extrados ties. *Int. Sem. Structural Analysis of Historical Constructions IV. Possibilities of numerical and experimental techniques,* eds. C. Modena, P. Roca & P.B. Lourenço, A. A. Balkema: London, pp. 745-752, 2004.

A study of site characteristics affecting seismic strengthening strategy

G. C. Yao & W. C. Chen
Department of Architecture, National Cheng Kung University, Taiwan

Abstract

This study demonstrates that geological characteristics will affect seismic retrofit strategy. An old temple had been retrofitted earlier but still suffered serious damage from recent earthquakes. In order to find a good solution to strengthen the temple, a computer structural model of the target building was developed and its accuracy was verified by using ambient vibration experiments. From nearby seismic station records, the strong motion characteristics were then studied using response spectrum analysis. The results indicated that increasing the building's stiffness would also increase the input energy. Therefore, another retrofit scheme was proposed to absorb earthquake energy instead of increasing the building's stiffness. A conceptual strengthening model was developed and the computer analysis was targeted for the temple's inter-story drift ratio to be kept under 0.5%. The effect of the final design was analyzed and discussed.
Keywords: retrofit, response spectrum analysis, inter-story drift ratio.

1 Introduction

Since the 1999 Chi-Chi Earthquake, higher requirements of seismic capacity have been increasingly recognized in Taiwan. However, many old buildings can only be retrofitted rather than rebuilt. There are plenty of retrofit methods. In addition to raising the stiffness of buildings to reduce the earthquake deformation [1–3], absorbing the input energy of earthquakes can also be used for retrofitting purposes [4]. In general, traditional retrofit methods are about adding shear walls and wing walls. Besides, the RC jacketing system and Steel-Framing System are also common in Taiwan. These seismic retrofit strategies are to raise the strength and stiffness of buildings to resist earthquakes and this is considered inexpensive and effective. But occasionally these methods cannot achieve the expected results if site conditions have special ground movement characteristics.

A case is considered in this study to demonstrate the importance of geological characteristics since they will affect seismic retrofit strategies. An old temple located in central Taiwan had been retrofitted earlier but still suffered damages from recent earthquakes. The authors' team was commissioned to help propose a feasible retrofit method.

2 Dynamic analysis

The old temple comprises two buildings. The front building is seven stories high, and the back one has three stories. The first and second stories of the two buildings are connected together. A computer structural model of the temple was built (fig. 1) and its accuracy verified by using an ambient vibration experiment.

Figure 1: Computer structural model.

The ambient vibration experiment was conducted to measure the vibration of the target building for 300 s. Acceleration sensors were placed on the 1st floor and the 7th floor to measure the ground motion and the building's vibration, respectively. The vibration measured on the 1st floor was considered as the input signal while that on the 7th floor was treated as the output signal. By using system identification, the frequency response function (FRF) could be obtained and thus the natural frequency of the temple was accurately found. Table 1 showed the natural frequency of the temple compared to the natural frequency of the computer structural model. The result indicated that the two were closely related, which meant that the computer structural model could actually simulate the structural behaviors of the target building. Moreover, the computer model could also provide an estimated response analysis with reliable accuracy.

Table 1: Fundamental natural frequency of the temple.

		East-west			North-south		
Ambient vibration experiment	Natural frequency	1.61Hz	1.56Hz	1.60Hz	1.61Hz	1.58Hz	1.63Hz
	Average value	1.59Hz			1.61Hz		
Computer structural model	Natural frequency	1.56Hz			1.62Hz		

After the accuracy of the model was confirmed, real earthquake records were used for computer dynamic analysis. First, according to the seismic codes, the design ground acceleration of the temple was calculated. Then the peak ground acceleration (PGA) of the selected earthquake records was adjusted to be exactly the same as the design ground acceleration. In comparison with all the other earthquake records, the 1999 Chi-Chi Earthquake was considered most representative and was chosen as the design earthquake. Comparing to the 2010 Jia-Shian Earthquake, the strongest ground motion happened in a very short interval (fig. 2) and the PGA was much stronger than the remaining ground acceleration. On the contrary, the Chi-Chi Earthquake lasted longer, and not just one peak ground motion but several strong ones also occurred and appeared on the record (fig. 3).

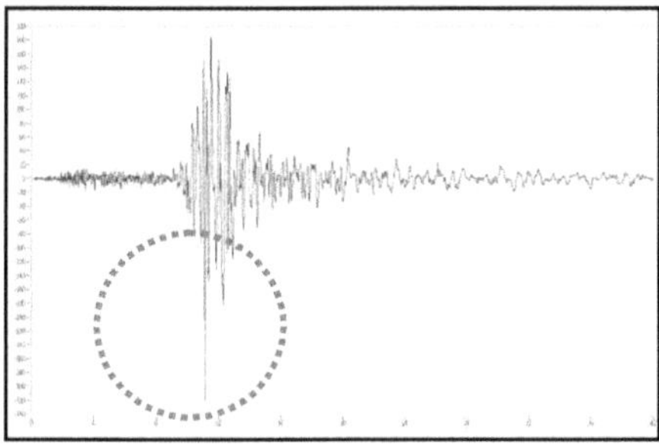

Figure 2: Jia-Shian earthquake record.

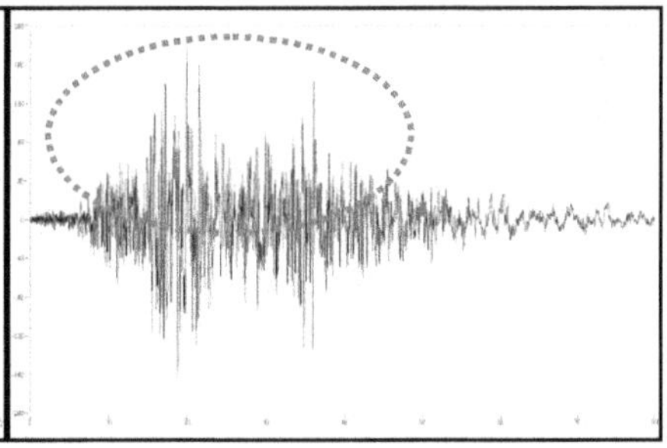

Figure 3: Chi-Chi earthquake record.

From the dynamic analysis, the temple's vibration behavior during earthquakes was predicted. Then the deformation of each floor was calculated to obtain the inter-story drift ratio shown in table 2. According to the seismic codes, a new building's inter-story drift ratio must be kept under 0.5%. It was found however that the 4 top floors of the front building and the 3rd floor of the back building swayed over this value. Furthermore, because the back building had many walls, which increased the local stiffness, the center of stiffness was close to the back building. Therefore, the temple was seriously affected by torsional excitation.

Table 2: The temple's inter-story drift ratio.

The inter-story drift ratio of the front building in east-west direction							
Floor	1st	2nd	3rd	4th	5th	6ht	7th
inter-story drift ratio (%)	0.08	0.37	0.24	0.57	0.67	0.65	0.60
The inter-story drift ratio of the front building in north-south direction							
Floor	1st	2nd	3rd	4th	5th	6ht	7th
inter-story drift ratio (%)	0.06	0.39	0.31	0.52	0.56	0.50	0.42
The inter-story drift ratio of the back building in east-west direction							
Floor	1st	2nd	3rd				
inter-story drift ratio (%)	0.09	0.35	0.21				
The inter-story drift ratio of the back building in north-south direction							
Floor	1st	2nd	3rd				
inter-story drift ratio (%)	0.05	0.24	0.85				

In this study, the temple's inter-story drift ratio was chosen to be the design criterion and the aim was to keep this ratio under 0.5% to maintain the elastic behavior of the structure. Another reason to select a design criterion instead of reviewing the stress in the structural elements was mainly because complete information about the construction materials and the plans of reinforcing bars could not be obtained. Besides, the temple was damaged in previous earthquakes and subsequently retrofitted without complete records being left. Therefore, all structural details were not available.

3 Earthquake records analysis

The dynamic analysis demonstrated the temple's vibration behaviors. However, it was believed that the geological characteristics around the building were vital to the generation of appropriate retrofit designs. Taipei area, for example, is a basin and its geology is soft. This geological characteristic will magnify the earthquake waves and produce a bigger ground motion. The building's vibration behavior is mainly influenced by the ground motion during earthquake. Therefore, in order to identify the geological characteristics around the temple, one seismic station nearest the temple (1.5 km), set by Central Weather Bureau, was chosen. The station recorded detailed data of each earthquake and among all the records three strong earthquake records (921 Chi-Chi, 1022 and 0304 Jia-Shian earthquakes) were selected to represent the ground movement characteristics of the area around the temple.

The earthquake records selected were characterised using the response spectrum analysis. In a response spectrum, the horizontal axis is the natural frequency of buildings, and the peak in the graph indicates the building frequency associated with the biggest energy during the earthquake.

Figs. 4 and 5 show the response spectra from the three earthquake records in two different directions (east-west and north-south). The dotted lines in the two figures indicate the natural frequency of the temple. From the curves in the figures, it is evident that if the natural frequency of the temple increases, more energy is transmitted to the building. The results send an important message;

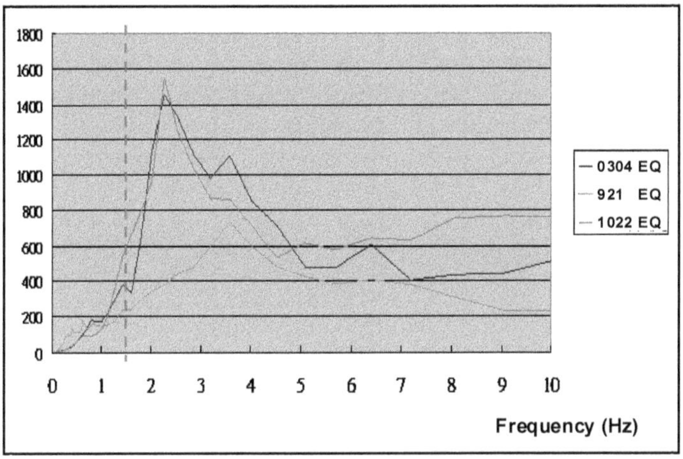

Figure 4: The response spectrum (east-west).

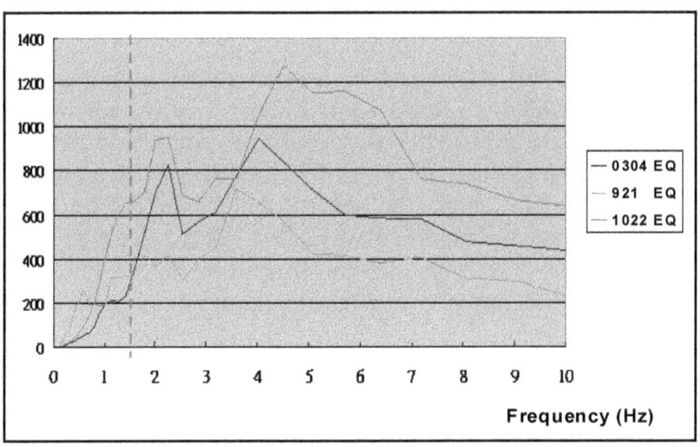

Figure 5: The response spectrum (north-south).

namely that the traditional retrofit methods of increasing the stiffness of a building are not appropriate for this case or, maybe, even worse. Buildings which had greater stiffness would also have bigger natural frequency. For this reason, if the traditional retrofit methods were followed, the stiffness of the temple would be increased and the natural frequency would also get higher.

4 Retrofit scheme

Because of the special ground movement characteristics at the temple, another retrofit scheme was proposed whereby earthquake energy was absorbed instead of the building's stiffness being increased. A viscous elastic material (VEM) damper [5] was used to achieve this goal. Its own stiffness can reduce the deformation of the buildings and its viscous property can provide damping to absorb the energy of the earthquakes. These advantages are appropriate in designing against earthquakes to raise the safety of buildings. And most importantly, a VEM damper will not increase the building's natural frequency in comparison with the traditional retrofit methods such as adding more columns or walls. This quality quite fits in with the special ground movement characteristics in this case. Table 3 shows the temple's natural frequency after being retrofitted compared to the original temple's natural frequency. The results make clear that the two were almost the same.

Table 3: The temple's natural frequency before and after retrofitting.

Temple's natural frequency	East-west	North-south
The temple's natural frequency before retrofitting	1.56Hz	1.62Hz
The temple's natural frequency after retrofitting	1.57Hz	1.62Hz

The Kelvin model [6] was used in the computer modeling of the VEM damper. The Kelvin Model is composed of a linear elastic element and a viscous damper connected in parallel. After verifying the computer input model of the VEM damper, 33 dampers were added to the computer structural model and the dynamic analysis was run again. The seismic effects would not be indefinitely decreased as the amount of damping increased. Thirty-three dampers was the most effective number after numerous modelling trials. In addition, the effectiveness of a VEM damper would be influenced by changing temperatures. The higher the temperature, the less effective a VEM damper is. As a result the analyses were conducted at temperatures of 15°C and 35°C. With the results from the computer time-history simulation shown in table 4, it is evident that, no matter how the temperature changes, the temple's inter-story drift ratios would all be under 0.5%.

Table 4: The temple's inter-story drift ratio before and after retrofitting.

The inter-story drift ratio of the front building in east-west direction (%)							
Floor	1st	2nd	3rd	4th	5th	6ht	7th
before retrofitting	0.08	0.37	0.24	0.57	0.67	0.65	0.60
after retrofitting (15°C)	0.06	0.23	0.16	0.39	0.47	0.44	0.41
after retrofitting (35°C)	0.07	0.25	0.17	0.41	0.48	0.46	0.44
The inter-story drift ratio of the front building in north-south direction (%)							
Floor	1st	2nd	3rd	4th	5th	6ht	7th
before retrofitting	0.06	0.39	0.31	0.52	0.56	0.50	0.42
after retrofitting (15°C)	0.06	0.38	0.30	0.43	0.46	0.45	0.40
after retrofitting (35°C)	0.06	0.39	0.31	0.45	0.48	0.46	0.41
The inter-story drift ratio of the back building in east-west direction (%)							
Floor	1st	2nd	3rd				
before retrofitting	0.09	0.35	0.21				
after retrofitting (15°C)	0.08	0.21	0.17				
after retrofitting (35°C)	0.08	0.23	0.18				
The inter-story drift ratio of the back building in north-south direction (%)							
Floor	1st	2nd	3rd				
before retrofitting	0.05	0.24	0.85				
after retrofitting (15°C)	0.05	0.23	0.42				
after retrofitting (35°C)	0.05	0.24	0.45				

In order to compare the traditional retrofit methods with the proposed scheme, the same number of concrete walls was added instead of VEM dampers. From table 5, it is evident that the temple's natural frequency increases significantly. According to the ground characteristics of the temple, this result would attract more energy into the building. Table 6 shows the results of the dynamic analysis when traditional retrofit methods are used. It is evident that the responses are disappointing. In most floors which needed to be improved, the inter-story drift ratios were still predicted higher than 0.5%.

Table 5: The temple's natural frequency with different retrofit methods.

Temple's natural frequency	East-west	North-south
temple's natural frequency before retrofitting	1.56Hz	1.62Hz
temple's natural frequency after retrofitting (VEM Damper)	1.57Hz	1.62Hz
temple's natural frequency after retrofitting (traditional method)	1.91Hz	1.98Hz

Table 6: The temple's inter-story drift ratio before and after being retrofitted.

The inter-story drift ratio of the front building in east-west direction (%)							
Floor	1st	2nd	3rd	4th	5th	6ht	7th
before retrofitting	0.08	0.37	0.24	0.57	0.67	0.65	0.60
after retrofitting	0.07	0.35	0.23	0.53	0.62	0.58	0.52
The inter-story drift ratio of the front building in north-south direction (%)							
Floor	1st	2nd	3rd	4th	5th	6ht	7th
before retrofitting	0.06	0.39	0.31	0.52	0.56	0.50	0.42
after retrofitting	0.06	0.38	0.31	0.48	0.53	0.48	0.40
The inter-story drift ratio of the back building in east-west direction (%)							
Floor	1st	2nd	3rd				
before retrofitting	0.09	0.35	0.21				
after retrofitting	0.08	0.29	0.19				
The inter-story drift ratio of the back building in north-south direction (%)							
Floor	1st	2nd	3rd				
before retrofitting	0.05	0.24	0.85				
after retrofitting	0.05	0.24	0.62				

5 Conclusion

This study discusses how the site characteristics affect a building's seismic strengthening strategy. An old temple had earlier been strengthened by traditional retrofit methods but still suffered serious damages from recent earthquakes. In order to find out why traditional methods were ineffective, the strong motion characteristics were studied by a response spectrum analysis. The results indicated that in this case, the traditional retrofit methods made the temple susceptible to a higher input of seismic energy during earthquakes.

As previously discussed, the traditional retrofit schemes were found not suitable for all cases. The reasons for this usually are differences in the geological characteristics. If a building in a certain area would not be subjected to higher earthquake energy when its natural frequency is increased, the traditional retrofit strategies are considered most proper. Their cost is less and the results are efficient. If, however, the situation is as in the present case, the traditional retrofit strategies are not the best choice anymore and energy absorption strategies are recommended. Therefore, if some research on the site characteristics of the target buildings could be carried out before performing a structural analysis, it would greatly help the selection of the most appropriate retrofit method.

References

[1] Tsai, Y.C., Huang, J.S., Tseng, Y.P. & Chen, W.C., *Detailed Evaluation of the Seismic Capacity of a School Building with Considering Strengthening*, National Center for Research on Earthquake Engineering, Taiwan, 2000).

[2] Chiou, Y.J., Shr, J.T., Shiau, F.P., Chiou, T.J., Ruan, J.P. & Huang S.J., *Field Test and Analysis for School Building Retrofitted by RC Jacketing System*, National Center for Research on Earthquake Engineering, Taiwan, 2008.

[3] Chiou, Y.J., Liou, Y.W., Huang, J.C., Shiau, F.P., Chiou, Y.J., Chiou T.J. & Huang, S.J., *Field Test and Analysis for School Building Retrofitted by Steel-Framing System*, National Center for Research on Earthquake Engineering, Taiwan, 2008.

[4] Advanced Civil Engineering Technology. *Seismic Retrofit Design of Advanced Civil Engineering Technology Factory Building*, 2010.

[5] Nippon Steel Corporation. *Seismic Technology – VEM Damper*, Nippon Steel Corporation, Japan, 2010.

[6] Computers and Structures, Inc. (2002). SAP2000 Analysis Reference Manual, Version 8.0. Computers and Structures, Inc., Berkeley.

Using seismic isolation systems for retrofitting historic buildings

I. Iskhakov & Y. Ribakov
Ariel University Center of Samaria, Ariel, Israel

Abstract

Historical buildings are usually characterised by relatively low height and their natural vibration period is rather low. Hence if such buildings are located in seismically active zones, using seismic isolation systems is a very effective way for improving earthquake resistance of these structures. Seismic isolation is well known, however its practical implementation only began in the last few decades. The main advantage of a seismic isolation system is that no structural elements should be added to the building in order to strengthen it. This is a very attractive benefit, being especially important for historical buildings. A traditional seismic isolation system is presented by the following two types – bearings and pendulums. The design of seismic isolation systems and selection of their dynamic properties depend on the features of the isolated structure that can be obtained by non-destructive testing. A method for structural dynamic parameters assessment was previously developed by the authors. For this reason an impulse test was carried out on a three-story structural part of a building. Sometimes the displacements at the isolation level are rather big and exceed the allowed limits. In such cases it is recommended to add dampers to the isolation system. Different types of variable friction dampers were proposed and tested by the authors. These dampers significantly reduce the displacements at the isolator level and also yield further improvement in the dynamic response of a building.

Keywords: historic buildings, seismic isolation, impulse tests, variable friction dampers, dynamic response.

1 Introduction

Seismic isolation is one of the significant developments in earthquake engineering in the past decades. It provides protection of structural elements and non-structural components without any need for addition of strengthening details that may change the original structure and cause irrecoverable damage to heritage architecture. In most cases buildings are supported on foundations, and have a fixed-base scheme. When a seismic isolation system is used, isolators can be installed between the building and its foundation or at any desired level of the structure. The isolators are flexible in the horizontal direction and reduce the dominant frequency of a building. As a result, the isolated building's response to an earthquake is not so strong like of the original one.

In most applications seismic isolators are installed between the foundation and the first floor columns (base isolation). In some cases isolators are added between the roof and the upper floor columns (roof isolation) [12]. It is possible to isolate any floor (for example, the open entrance floor in typical residential buildings in the Mediterranean area and other countries with similar specific climate conditions).

The first application of a base isolation system was in New Zealand in 1974, in the US – in 1984 and in Japan – in 1985. Currently there are more than 80 base isolated buildings and 150 bridges in the US, 1000 buildings and 500 bridges in Japan and numerous applications in other countries. For example, in the USA almost 55% of government buildings and 45% private buildings are base isolated. Among them 12 historic buildings are retrofitted.

As known, the main advantage of using seismic isolation systems for protection of historic buildings is that the retrofitting works can be performed at the basement or roof levels. In both cases the addition of isolation results in no changes in the structural system and saves the historic value of the heritage building. An additional benefit of this solution is its relatively low cost compared to other retrofitting techniques.

The main idea of seismic isolation is "decoupling" the building from the ground motion or dividing it into two subsystems. If base isolation is used, the transmission of seismic energy to the building is significantly reduced by increasing the isolated building's dominant period, allowing the building to displace with rather small inter-story drifts. Adding isolators at the roof level yields rather lower increase of the dominant period, compared to base isolation. However, in this case damping in the system "building – roof" increases significantly. Both systems allow decreasing internal forces in the structural elements of the buildings during earthquakes. Classic base isolation systems are usually obtained by using two main isolator types – bearings (fig. 1) or pendulums (fig. 2).

As can be understood from fig. 1, a bearing is a sandwich, obtained by alternating steel plates and rubber layers, vulcanized to form a single unit. As shown in fig. 2, a pendulum consists of a concave plate, an articulating slider and a housing plate. Both types of isolators are able to displace horizontally in any direction from the centre depending on the seismic excitation's direction.

This paper focuses on reviewing the application of seismic isolation systems to historic buildings, non-destructive impulse testing for measuring structural dynamic parameters and the use of variable friction dampers for reducing displacements at the base isolator's level.

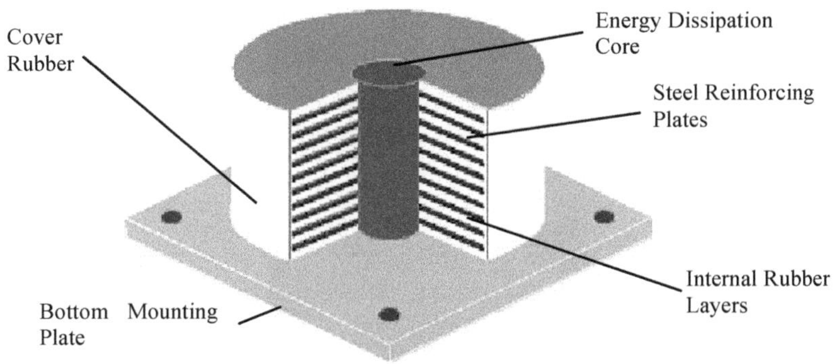

(a)

(b)

Figure 1: Elastomeric bearings: (a) general scheme, (b) a column with a base isolator [1].

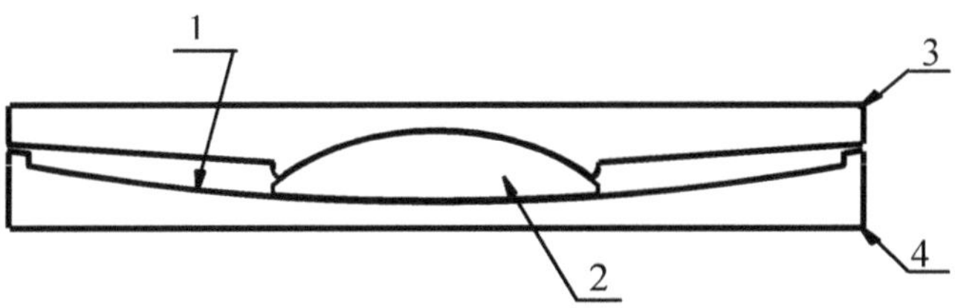

Figure 2: Pendulum type isolator: 1 – concave surface, 2 – convex slider, 3 – housing plate, 4 – concave plate.

2 Seismic isolation for historic buildings

2.1 Oakland City Hall

The building was constructed in 1914 as the first high-rise government office in the United States and remains one of the country's finest examples of Beaux Arts architecture. The building suffered severe damage during the 1989 Loma Prieta earthquake and was immediately closed to occupancy. It is a beautiful building, with its three-story central rotunda, 10-story base, and highly decorative clock tower. After the earthquake, about 20% of the building's lateral bearing capacity

Figure 3: Base isolated Oakland City Hall.

was lost in the north-south direction and about 30% in the east–west one. Extensive cracking of the structural elements appeared.

It was decided to retrofit the structure using a base isolation system. The main aim was to restore the building to its majestic self and to increase its seismic resistance, so that future earthquakes would not cause similar damage. The building is now one of the tallest base-isolated structures in the world. It has a complete steel frame. The floors and roofs are made of reinforced concrete slabs supported on the frame. The foundation system is a reinforced concrete plate over the entire basement area. 112 isolators were installed in the basement.

2.2 San Francisco City Hall

The San Francisco's Old City Hall (fig. 4(a)) was constructed at the end of the XIX century. It was one of the finest examples of French Renaissance architecture in the country. During the 1906 earthquake, the building was totally damaged; only the dome remained intact. It was finally rebuilt in 1915 (fig. 4(b)). During the 1989 Loma Prieta earthquake, the City Hall was again subjected to strong seismic excitation. The dome twisted on its steel frame like a cap on a bottle; cracks in the walls and concrete floor slabs occurred at all levels [3]. The building was not in danger of collapse, but, according to expert opinion, its seismic resistance was not enough to withstand the next big earthquake.

About ten years ago the building was isolated from its foundation and put on 530 bearings, which were designed to dissipate earthquake energy and allow the building to move horizontally up to 65 cm without shaking (fig. 4(c)). To install the isolators, the building columns were jacked up and cut, and the isolators were located between the columns and the foundation. Today, this structure is one of the finest examples of French Renaissance architecture and known as one of the most seismically safe buildings.

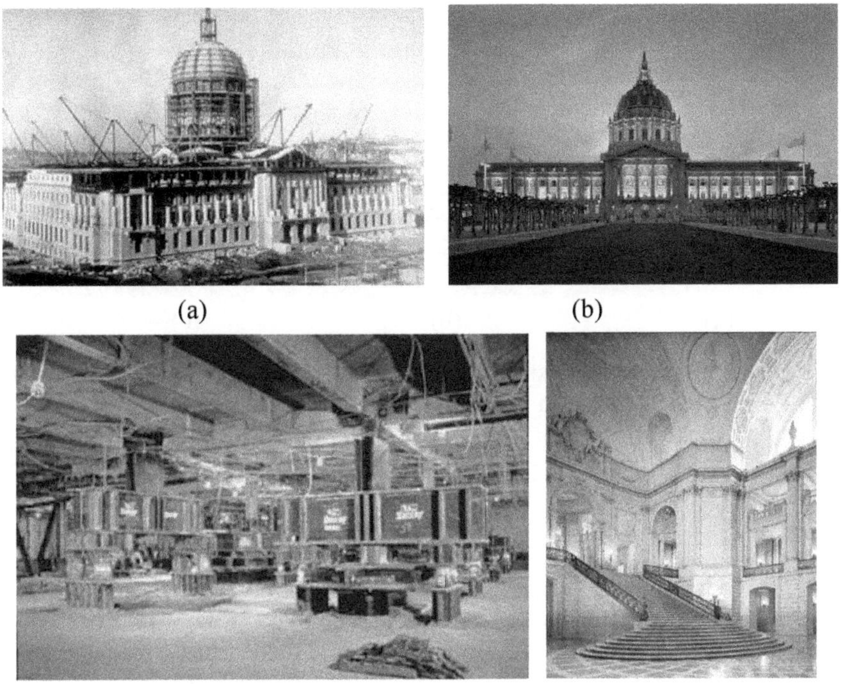

(a) (b) (c) (d)

Figure 4: San Francisco City Hall: (a) before reconstruction, (b) modern view, (c) installing isolators under the columns, (d) entrance hall.

2.3 Base isolation of historic buildings in New Zealand

One of New Zealand's most well known landmark buildings has undergone a complete seismic renovation [4]. The Parliament Buildings (fig. 5(a)) were originally built in the 1900s and located approximately 400 m from one of the region's major fault lines. Over 400 base isolation elastomeric rubber bearings were installed between the structure and its foundation. The isolators allow the building to move horizontally by about 40 cm during an earthquake.

The Old Bank of New Zeeland (fig. 5(b)) was reconstructed and provided with a base isolation system. This process involved tying together four historic buildings to create an integrated retail and office development [4].

The Museum of New Zealand is home to some of New Zealand's national artefacts and has a rich knowledge of the country's past (fig. 5(c)). It is the largest in the Southern Hemisphere base isolated structure weighting about 64 000 ton [4]. The base isolation system allows the building to move horizontally up to 50 cm. The isolated building is designed to withstand an earthquake of 7.5 Richter scale magnitude.

However, none of the heritage structures, retrofitted with base isolation systems, have been subjected yet to a strong earthquake, but other base isolated

(a) (b) (c)

Figure 5: Base isolated structures in New Zeeland: (a) Parliament building, (b) Old bank in Wellington, (c) the Museum of New Zealand.

buildings showed enhanced behavior in this case (like the USC Hospital in California [8]).

2.4 Roof isolation systems

The effectiveness of roof isolation for the aforementioned 13-story building was studied [13]. The building is located in Sherman Oaks, California and designed around 1964. It has 13 stories, 2 basements, a pile foundation, a rectangular plan 22 m wide and 58 m long. It presents a moment-resisting RC framed structure with seven bays along its longitudinal direction and two along its transverse one.

The building was analyzed with a roof isolation system, including isolation bearings and viscous dampers. It was reported [13] that the roof isolation system keeps all members within their linear elastic range and all inter-story drifts within the allowable limit. On average, the roof isolation system reduces the inter-story drifts by 35 to 40% in each of the building's directions.

Variable friction dampers (VFD) that are discussed in detail below were also proposed to be used as a part of a roof isolation system [14]. A method for selection of damper parameters was developed. It was shown that using roof isolation allows significant reduction in structural response. Absolute peak floors displacements were reduced by up to 50% in a structure with roof isolation system compared to an uncontrolled one.

Energy, dissipated in the roof isolation system with VFD was similar to that in a system with viscous dampers. There was no significant change in the peak base shear forces of the structure. Hence roof isolation systems may be effectively used for retrofitting heritage buildings.

3 Experimental methods for the estimation of dynamic parameters of heritage buildings

The selection of base isolation systems parameters directly depends on the structural dynamic characteristics. Known theoretical methods cannot be used for calculation of these characteristics because of the following reasons: the materials properties, element masses, static schemes for old buildings and the hysteretic properties of their elements are not known exactly; it is difficult to take into account all the damages caused to the building during its operation.

One of widely known and often used methods is an impulse test (described below). It is effective for structural systems with concentrated masses. However, this is not the case for irregular structures or structures with distributed masses. Masonry and stone heritage buildings belong to these categories. In this case other methods for estimation of structural dynamic parameters should be applied.

One such method is a micro-seismic vibration due to a low power underground blast at a certain distance from the building. The blast load capacity and the distance from the structure in this case are calculated to avoid inelastic structural behaviour [5]. Another way to apply vibrations to the building is to put a vibration machine at the basement level. The parameters and the regime of the machine are regulated to cause only linear structural vibrations [6].

If the building is located in a region with high seismic activity, it is possible to provide it with seismometers working in a waiting regime. The seismometers may be tuned to be activated when even a low-magnitude earthquake would occur [6]. For framed heritage buildings with relatively low height and rather high flexibility, an impulse testing method developed by the authors [7] may be used.

4 Impulse test procedure

According to this method, the tests are carried out on a framed building. For example, tests, performed on a building part consisting of two columns (each three floors high) and three slabs, were described in a previous work [7]. The columns were fixed in sockets on a 30-cm thick raft foundation. Each of the slabs was constructed of two pre-cast over-column plates and a between-column plate 16 cm thick (fig. 6). The overall dimensions of the part were 3×9 m, the height of each story being 3 m.

Dynamical impulse loads were applied to the column-slab joint at the roof level in the plane of the frame. The impulse was applied by instantaneous application of the load by means of a guy cable (fig. 7), which included a bar made of high-strength steel. The forces were controlled by rupture of a steel bar and were equal to the required impulse load. Under the impulse load, the tested frame vibrated freely. A steel cable and a hook were installed across the top of the third-floor column. The cable included a wire for pulling the frame. One end of the wire was tied to the hook and the other end to a bulldozer, which tensioned the wire until its failure.

In order to obtain the dynamic parameters of the structure, two accelerometers were located horizontally in the plane of the investigated frame at each floor, and also on the ground. The accelerometers were calibrated and tuned using a laboratory shaking table prior to the start of the tests. The bars were also subjected to preliminary testing in the laboratory in order to measure their absolute deformation under the rupture forces. These measured deformations corresponded to the maximum displacements of the building's roof story under the impulse loads applied.

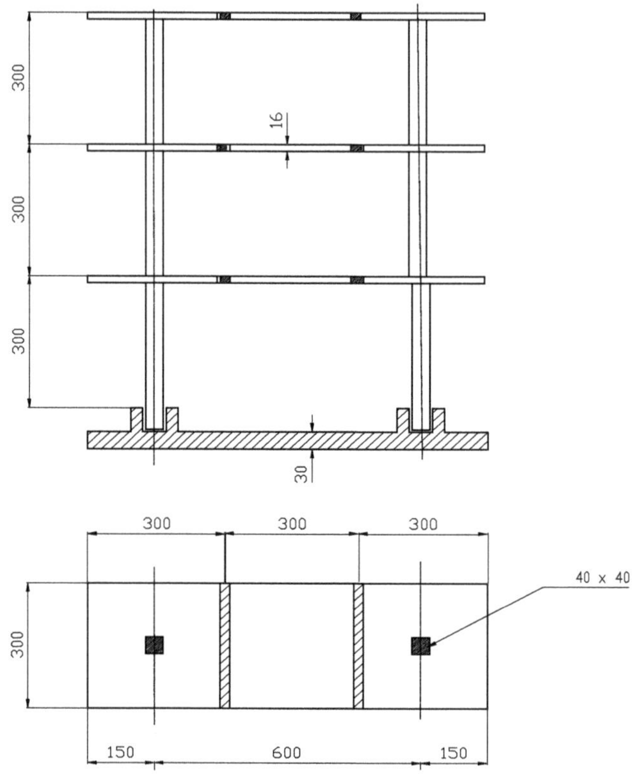

Figure 6: Structural scheme of tested part (following [7]).

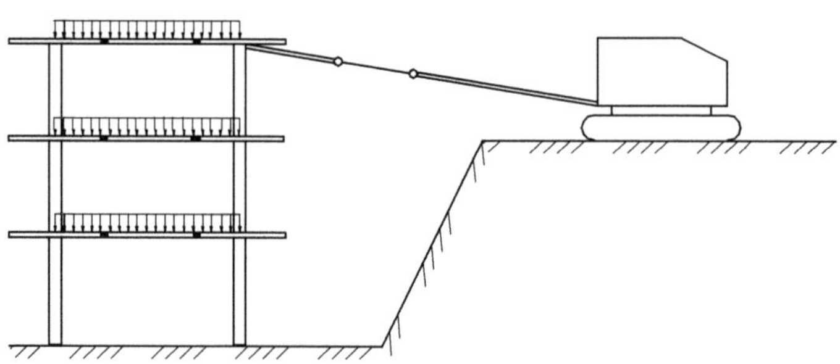

Figure 7: Application of impulse to building part (following [7]).

The acceleration records were then analyzed using the MATLAB software. The natural periods were determined by computation of a frequency response curve showing the resonant peak corresponding to each natural frequency of the structure [7]. From these data the real damping ratio of the first vibration mode was determined using a logarithmic decrement.

In the above described tests, the bar's section area was selected to vary the impulse forces applied to the structure. This approach may be also used in testing heritage buildings in order to avoid any damage to its elements. It should be noted that the impulse force magnitude should be enough to yield the structural vibrations required to obtain the buildings dynamic parameters.

5 Using dampers in seismic isolation systems

As mentioned above, seismic isolation systems significantly reduce inter-story drifts, seismic forces and inelastic deformations in structural elements, preventing damages that may be caused to a building during an earthquake. Heritage architecture is especially sensitive to these influences. However, in some cases the influences are still unacceptable. Moreover, using base isolation systems yields relatively big displacements at the isolators' level, which is not always desired.

Addition of dampers to base and roof isolation systems yields further improvement in structural response to earthquakes and decreases the displacements at the isolators [8]. Passive fluid viscous, friction and other types of dampers may successfully be used for this reason [9, 13]. It is known that buildings with friction dampers do not return to the initial position after the earthquake. In order to overcome this drawback variable friction dampers were developed and tested [10, 11].

A variable friction damper (fig. 8) consists of a tube (1), a wedge (2), elastic strip elements (3) and a connection clip (4). The wedge may be linear or curved. It can move ahead and back along its axis. The strips are fixed on the tube by the connection clip, forming an elastic strip system. The stiffness of this system may be regulated by changing the location of the connection clip along the tube. The free ends of the cantilever strips have a contact with an inclined surface of the wedge.

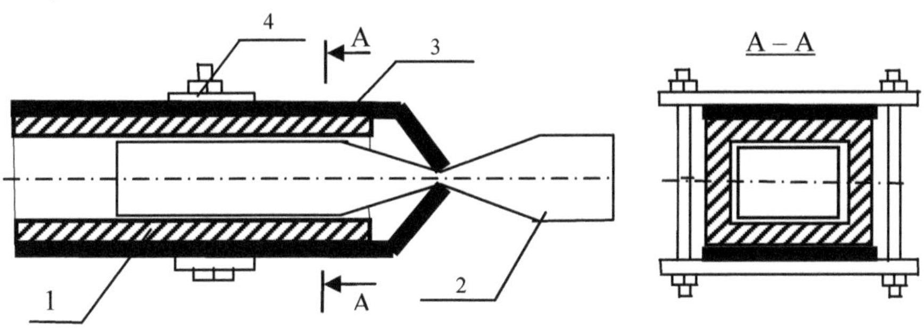

Figure 8: A principal scheme of the damper: 1 – square section tube; 2 – wedge; 3 – elastic strip elements; 4 – bolted connection clip (following [10]).

The dampers' mechanical properties and hysteretic behaviour were studied theoretically and verified experimentally using laboratory models [10, 11]. These dampers may be incorporated as part of seismic isolation systems in order to yield enhanced structural response and to limit the displacements at the isolators.

6 Conclusions

Seismic isolation has significant benefits for the earthquake protection of historic structures. It provides high protection level after a strong earthquake by reducing the buildings' seismic response. Thus, seismic isolation is an acceptable and effective way for the protection of historic buildings in areas of high seismic activity.

Recent earthquakes have shown that isolated historic buildings retrofitted to withstand earthquakes behaved better compared to fixed-base ones. The main advantage of a seismic isolation system is that no elements should be added to an isolated building, which is especially important for heritage architecture.

Design of seismic isolation systems and selection of their properties usually depend on the dynamic characteristics of the isolated building. The latter can be obtained by non-destructive impulse testing of the structure before its protection.

Rather big displacements at the isolators' level can be reduced by adding supplemental dampers to the isolation system. Effective variable friction dampers were proposed and tested by the authors. The effectiveness of these dampers in roof isolation systems was proved numerically. It was demonstrated that these dampers significantly reduce the displacements in the isolators and also yield further improvement in the seismic response of a building.

References

[1] Los Angeles City Hall seismic rehabilitation project, Internet resource http://www.lacityhall.org/isohow.html

[2] Zayas, V., Friction pendulum seismic isolation bearings, Internet resource http://www.cif.org/Nom2000/Nom08_00.PDF

[3] San Francisco City Hall, Internet resource http://www.celebratingeqsafety.com/sf-city-hall

[4] Holmes Consulting Group – Base Isolation, Internet resource http://www.holmesgroup.com/baseisolation.html

[5] Negmatulaev, S.H., Djabarov. M.D. & Sirma, O.V., Full-scale dynamic tests of clay brick buildings using seismic-blast method. *Proc. of the Int. Conf. on Modern Aspects in Development of Seismic Resistance and Seismology:* Dushanbe, pp. 172-178, 2005 (in Russian).

[6] Iskhakov, I., *Calculation and Design of RC Shells for Seismic Regions*, Maorif: Dushanbe, 1983 (in Russian).

[7] Iskhakov, I. & Ribakov, Y., Selecting the properties of a base isolation system based on impulse testing of a three-story structural part. *European Earthquake Engineering,* **19(1)**, pp. 38-42, 2005.

[8] Naeim, F. & Kelly, J., *Design of seismic isolated structures: from theory to practice*, Wiley: NY, 1999.

[9] Chang, S., Makris, N., Whittaker, A. & Thompson, A., Experimental and analytical studies on the performance of hybrid isolation systems. *Earthquake Engineering and Structural Dynamics,* **31**, pp. 421-443, 2002.

[10] Ribakov, Y., Blostotsky, B. & Iskhakov, I., Theoretical model and laboratory tests of variable friction damper. *European Earthquake Engineering*, **1**, pp. 36–42, 2006.

[11] Ribakov, Y., Blostotsky, B. & Iskhakov, I., An efficient variable friction damper with a curved wedge. *European Earthquake Engineering*, **2**, pp. 32-38, 2006.

[12] Villaverde, R., Roof isolation system to reduce the seismic response of buildings: a preliminary assessment. *Earthquake Spectra*, **14**, pp. 521-532, 1998.

[13] Villaverde, R., Implementation study of a seismic roof isolation system in 13-story building. *Journal of Seismology and Earthquake Engineering,* **2(2)**, pp. 17-27, 2000.

[14] Ribakov, Y. & Agranovich, G., A roof isolation system with variable friction dampers for seismic protection of multistory buildings. *European Earthquake Engineering,* **21(1)**, pp. 10-15, 2007.

Numerical simulation of the dynamic and earthquake behavior of Greek post-Byzantine churches with and without base isolation

G. C. Manos & N. Karamitsios
Aristotle University of Thessaloniki, Greece

Abstract

The dynamic and earthquake response of post-Byzantine stone masonry churches, subjected to a damaging earthquake sequence in the region of Western Macedonia in Greece, is examined. One of these "Basilica" form churches is analyzed numerically. The study extends to the behavior of a two-story box-type church with a simple orthogonal vault and a wooden roof, which belongs to an 11th century monastic-complex. These two structures are also combined and analyzed with a base isolation system. The predicted performance of the various structural elements, located at the peripheral masonry walls, is then checked by applying an assumed Mohr-Coulomb failure envelope that is believed to represent the limit-state stone-masonry behavior. The conceptual introduction of a base-isolation system was quite effective in lowering the demands and thus resulting in Capacity/Demand ratio values larger than one, which is assumed to indicate acceptable structural performance. However, the actual introduction of base isolation under an existing old stone-masonry structure faces considerable difficulties, despite the favorable results of the numerical modeling. For "The Virgin Mary of Tourniki" these difficulties must be addressed because this old church has been moved to a different location uphill due to flooding. Thus, introducing a base isolation scheme under this structure at its new location is a feasible solution in this particular case with no significant additional cost.

Keywords: Byzantine churches, stone masonry, dynamic response, earthquake performance, base isolation.

1 Introduction

During the last thirty years, various parts of Greece have been subjected to a number of damaging earthquakes ranging from M_s = 5.2 to 7.2 on the Richter scale [1]. One of the most demanding tasks for counteracting the consequences of these seismic events has been to ensure the structural integrity of old churches, built in periods dating back to 400 A.D.; in many cases, these structures sustained considerable damage [1, 4, 5]. In what follows, selected results and summary observations of the dynamic and seismic behavior of two types of structures that characterize a considerable number of churches belonging to the so-called post-Byzantine period (16th to 19th century A.D.) are presented. The first structure is of "Basilica" form and has a number of variations in plan and elevation whereas the second type is a simple box-type structure. This study examines the seismic behavior of one church belonging to the 1st structural type (typical post-Byzantine Basilica), representing at least five churches damaged by an earthquake sequence, and another church belonging to the 2nd structural type (Virgin Mary at Tourniki). These churches are located in the region of Western Macedonia in Greece and they were all subjected to the damaging strong Kozani earthquake sequence of May, 1995 (fig. 1) [2, 3, 7].

Figure 1: Map of the epicentral area with the location of the damaged churches.

The numerical results presented here are part of an attempt to numerically explain the observed behavior. The investigation includes modal analysis results but no attempt was made to compare these results with in-situ measurements for dynamic characterization. For the examined structures, the numerical analysis predicts the demands imposed on the masonry elements by critical load combinations which include earthquake loading as specified by the Greek Seismic Code [9]. The performance of the various masonry structural elements is evaluated in section 4 by examining the values of the Capacity/Demand ratio. The capacities are obtained by assumed strength values, without the incorporation of any safety factors. In general, the predicted damage regions agree well with observed damage [4, 6, 7]. It is obvious that an important next step is the rehabilitation of the damaged masonry. The structural repairs of the stone-masonry churches examined here are inhibited by restrictions arising from retaining the visible stone-masonry architectural forms on the exterior and the frescos on the interior of the masonry walls. An acceptable type of repair is the application of low-pressure lime-mortar injections enriched with ceramic powder, resulting in a modest increase in the capacities of the various structural elements. The application of techniques used for the repair of reinforced concrete structural elements (jacketing or use of fiber reinforcing plastics), providing considerable capacity increase, is not acceptable for the stone masonry of such cultural heritage structures. An alternative solution is lowering the demands by the use of base-isolation schemes as the ones investigated here. However, the introduction of base isolation under an existing old stone-masonry structure also faces considerable difficulties, despite the advantages indicated by the numerical results presented here.

1.1 Typical post Byzantine Basilica

The first case is a three-nave post Byzantine "Basilica" structural form, typical overall of a number of churches with similar geometry (fig. 2), which were damaged by a strong earthquake sequence in the region of Western Macedonia, Greece during the Kozani Earthquake of 1995 (fig. 1). These churches were located at various villages in the epicentral area (Lefkopigi, Paleochori, Knidi, Sarakina, Panagia (fig. 1)).

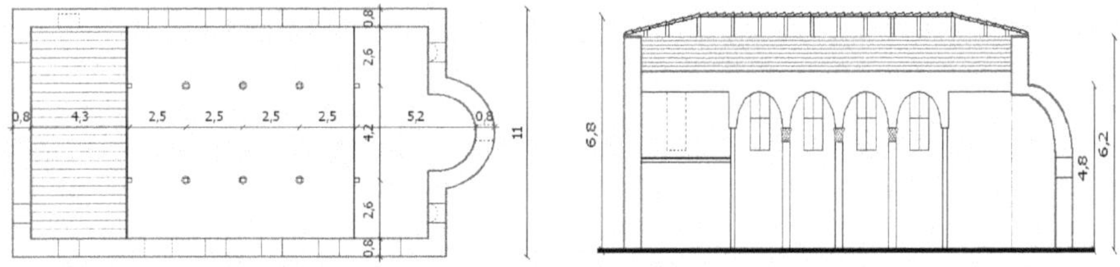

Plan

Cross section

Figure 2: Typical post-Byzantine "Basilica".

The overall dimensions of this "typical" case are 19 m length, 11 m width and 6.8 m height of the central nave (from ground level to the top of the roof). The internal colonnades are made of wood (fig. 2).

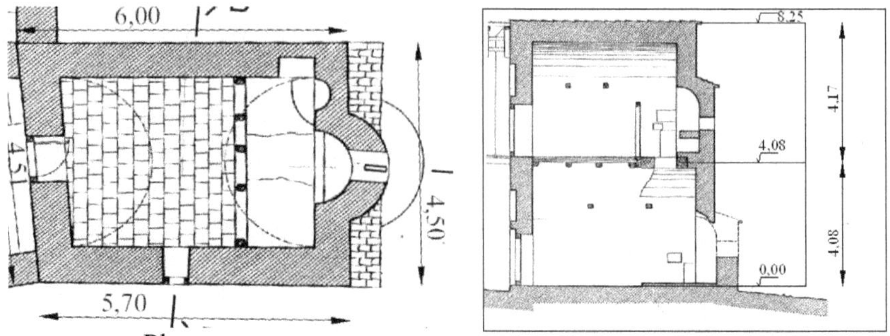

Plan

Cross section

Figure 3: The church of Virgin Mary at the monastic complex of Tourniki.

1.2 The Virgin Mary of Tourniki

This is a simple box-type structure of rectangular shape, formed by the peripheral walls; a semi-cylindrical apse is part of the East wall for both the 1st and the 2nd story. This is the simplest form of a "Basilica", that is, a single-space rectangular structure without interior separations. This two-story box-type Basilica is such a structure (fig. 3), which is devoted to the Virgin Mary and belongs to the monastic complex of Tourniki; it is a rather rare structural system that can be found in the monastic complexes of Mt. Athos. The behavior of this two-story Basilica is numerically simulated by focusing on the super-structure and in particular on the influence of the cylindrical vaulting system that exists at the 1st story level as well as at the roof. The numerical investigation for this second church is also extended to the effect of introducing a base isolation system.

1.2.1 Main features of the church

The establishment of the monastic complex and the construction of this church is set at the 11th century A.D. This church is a two-story box-type "Basilica" structural system formed by two longitudinal walls (North and South) and two transverse walls (East and West) with a thickness of 700 mm. The East transverse wall is dominated by two apses (one for each story) with a thickness of 700 mm. A relatively complex system is constructed at the East transverse wall with the two apses in order to transfer the gravitational forces of the top East wall and apse, which are offset by 700 mm towards the interior of the church, when compared with the East wall and apse of the bottom story (fig. 3). The top story ceiling is formed by a longitudinal semi-cylindrical vault with a thickness of 350 mm, which supports the wooden roof together with the peripheral walls that extend appropriately for this purpose. Despite some small differences in the length of the North and South longitudinal walls, the structural system as a whole can be assumed to retain a mid-axis symmetry with respect to the East-West longitudinal direction. Apart from the difference in the length of the longitudinal walls, the exact location of the window openings also deviates from this symmetry; however, these deviations are not believed to be very

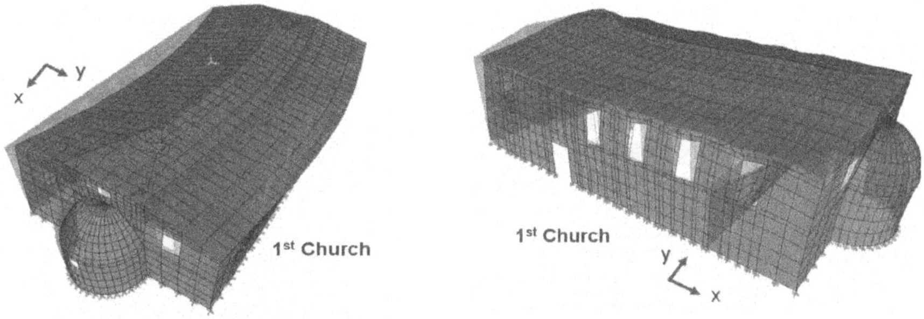

Translational y-y
$T_y = 0.102$ s, $u_y = 51.38\%$
(Non-isolated)

Translational x-x
$T_x = 0.069$ s, $u_x = 49.63\%$
(Non-isolated)

Figure 4: Dynamic characteristics of the "*Typical Basilica*".

significant. This church was subjected to the same 1995 strong earthquake (fig. 1) when the monastic complex was not in use; damage visible at the cylindrical vaults cannot be linked directly to this event.

2 Modal analysis of the post-Byzantine basilicas

A linear-elastic modal analysis was conducted assuming a value for the Young's modulus for the masonry walls equal to 2500 MPa. The mass density of these stone masonry walls was assumed to be equal to 2.70 t/m^3. All walls were meshed into shell finite elements. The arches on top of the internal colonnades as well as the wooden roof were also modelled; the Young's modulus of all wooden parts was taken equal to 8400 MPa and the corresponding density 0.66 t/m^3.

2.1 "Typical Post Byzantine Basilica"

The translational eigen-mode in the transverse North-South (y-y) direction is the one with the longest eigen-period (fig. 4, $T = 0.102$ s). The structural response in this mode displaces the longitudinal peripheral walls mainly out-of-plane; this is done with the transverse peripheral walls resisting mainly in-plane. The translational eigen-mode in the longitudinal East-West (x-x) direction is the next longest eigen-period (fig. 4, $T = 0.069$ s). The structural response in this mode displaces the longitudinal peripheral walls mainly in-plane; this is done with the transverse peripheral walls resisting mainly out-of-plane. Each one of these modes is associated with approximately 50% of the total mass of the structure. Higher order modes are associated with relatively small portions of the total mass.

2.2 "The Virgin Mary of Tourniki"

The first two horizontal translational mode shapes and eigen-periods are shown in fig. 5. As can be seen, due to the two-story formation, these two eigen-periods are relatively longer than the corresponding values of the previously examined typical post-Byzantine basilica. Moreover, these two eigen-modes are also associated with much larger modal mass ratios than the corresponding values of the previously examined basilica, a fact that must be attributed to the box-type form of this structure; moreover, the peripheral masonry walls have relatively small dimensions to be excited in an out-of-plane manner.

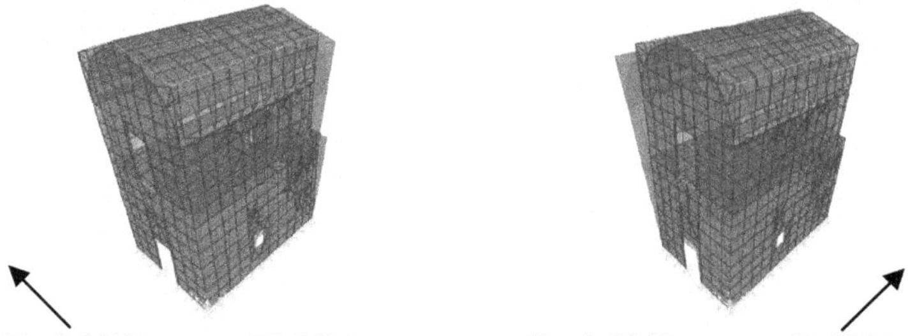

$T = 0.1375$ s, $u_y = 73.94\%$
Non-isolated

$T = 0.0942$ s, $u_x = 68.02\%$
Non-isolated

Figure 5: Dynamic characteristics of the "Virgin Mary of Tourniki".

3 Demands from gravitational and earthquake loading

The behavior of the selected structures was examined next when they were subjected to three distinct loading conditions. The forces for all these three loading conditions were applied in a static manner. The structures were assumed to have all three translational degrees of freedom (u_x, u_y, u_z) restrained for all masonry at the foundation level. The first loading case included the dead (G) loads of all parts plus the live (Q) loads (mainly snow at the roof level plus the live load at the level used as women's quarters). For the second and third loading conditions, the earthquake forces E_x and E_y were applied along the *x-x* and the *y-y* axis, respectively. This was done in a simple way assuming acceleration for all the parts of the structure equal to 1*g* (where *g* is the acceleration due to gravity). The dynamic nature of the seismic forces was taken into account in a separate series of simulations presented in the next section.

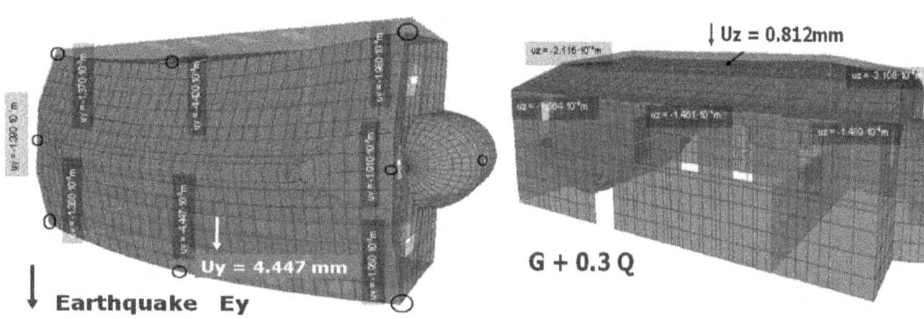

(a) (b)

Figure 6: (a) Deformations due to E_y. (b) Deformations due to gravity.

3.1 "Typical Post Byzantine Basilica"

The results for the Post-Byzantine Basilica are shown in figs. 6(a) and (b). As can be seen in these figures, this structural system is more flexible in the transverse than in the longitudinal direction. The resistance of the internal colonnades to either the *x-x* or the *y-y* seismic forces is very small as these structural elements are quite flexible. The maximum horizontal displacement at the roof level is equal to 1.94 mm for the loading case E_x whereas it attains a value of 4.447 mm for the loading case E_y (fig. 6a), more than double than the maximum horizontal (*x-x*) displacement. The seismic forces are mainly resisted by the in-plane action of the peripheral walls parallel to the direction of these forces as well as by the out-of-plane action of the peripheral walls normal to the direction of these forces. The maximum value of deformations due to the gravitational forces is equal to 0.812 mm (fig. 6b); this occurs along the vertical direction at mid-span of the top of the roof. The vertical deformations at the top of the peripheral walls vary from 0.1 mm to 0.2 mm; moreover, the out-of-plane flexibility of the longitudinal walls results, at their top, in out-of-plane deformations of the order of 0.15 mm when the structure is subjected to the gravitational forces.

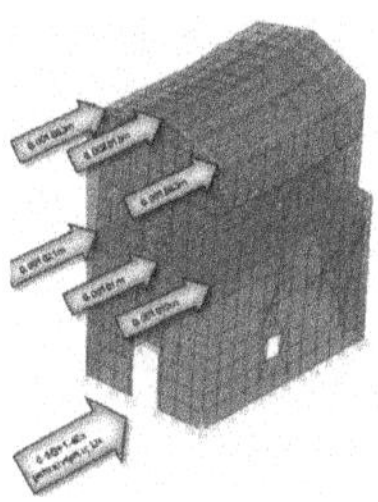

Non-Isolated

$0.9G + 1.4E_x$
$\delta_{xmax} = 2.015$ mm
(top of the roof)

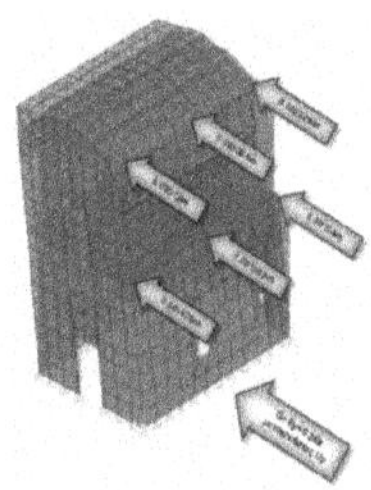

Non-Isolated

$0.9G + 1.4E_y$
$\delta_{ymax} = 3.081$ mm
(top of the roof)

Figure 7: Displacements of "The Virgin Mary of Tourniki".

3.2 "The Virgin Mary of Tourniki"

The small size in plan for this church and the box-type formation results in relatively small horizontal displacement amplitudes at the roof level for the combination of gravity loads and seismic forces along the *x-x* and the *y-y* direction as depicted in figure 7. The maximum value of the horizontal displacements at this level is approximately 3.1 mm and develops, as expected, along the direction with the relatively smaller stiffness (North-South, *y-y*). This structural form is designated in this figure as non-isolated to differentiate it from the same structure being isolated, as presented in section 5.

4 Stress results

This time the design spectrum of the Greek Seismic Code (Greek Seismic Code 2000, ref. [9]) was utilized for seismic zone I (ground design acceleration 0.16*g*), soil category B, response modification factor $q = 1.5$ and importance factor 1.3.

In the spectral dynamic analyses that were conducted, the resultant seismic forces were obtained from the Greek Seismic Code response spectrum and the following loading combinations (G the dead loads, E_x and E_y the earthquake action in the x and y directions): $0.9G+1.4E_y$ / $0.9G+1.4E_x$ / $G+E_y+0.3E_x$ / $G+E_x+0.3E_y$. From all load combinations, the most critical in-plane demand values, either normal or shear stresses, can be identified for all four peripheral walls. This can also be done for the most critical out-of-plane normal stress demand values for all four peripheral walls. Next, certain commonly used masonry limit-state criteria were adopted for either in-plane tension-compression or shear or out-of-plane tension.

Table 1 lists values which were assumed to be valid critical mechanical properties for the masonry segments [6, 10, 11]. Moreover, a Mohr-Coulomb failure envelope was adopted for the in-plane shear limit state of the stone masonry, when a normal stress σ_n is acting simultaneously; σ_n is defined through the relationship

$$f_{vk} = f_{vko} + 0.4\sigma_n \tag{1}$$

where f_{vko} is the shear strength of the stone masonry when the normal stress is zero; f_{vko} was assumed to be equal to 0.192 N/mm^2.

Table 1: Assumed mechanical characteristics of the stone masonry.

	Stone masonry compressive strength (N/mm^2)	Stone masonry tensile strength (N/mm^2)	Shear strength f_{vko} (N/mm^2)
Upper limit	3.846	0.250	0.192
Lower limit	1.00	0.192	0.192

All masonry parts of the three studied structures were examined in terms of in-plane and out-of-plane stress demands posed by the considered load combinations against the corresponding capacities, as these capacities were obtained by applying the Mohr-Coulomb criterion (1) or the upper stone masonry compressive and tensile strength limits listed in Table 1.

Such results are presented in detail elsewhere [6–8]. Selective results obtained from this evaluation process are shown in figs. 9 (*typical Basilica*) and 10 (*Virgin Mary of Tourniki*). R_σ and R_τ represent, respectively, the ratios of the in-plane tensile and shear strength values to the corresponding demands whereas R_M represents the ratio of the out-of-plane tensile strength value to the corresponding demand. Ratio values smaller than 1.0 predict a corresponding limit-state condition leading to damage. As can be seen from the capacity over demand ratio values presented in figs. 9 and 10, there are regions in the masonry walls where these ratio values are well below 1, indicating damage. All these predicted zones of potential failure are credible, as can be deduced from such damage patterns observed in the two studied structural systems after the 1995 strong earthquake event (fig. 1 and fig. 8).

(a) (b)

Figure 8: (a) Typical damage of a “Basilica” at the top of the south longitudinal wall as well as of the east transverse wall with an apse (Leykopigi). (b) Typical damage of a “Basilica” at the west transverse wall (Knidi).

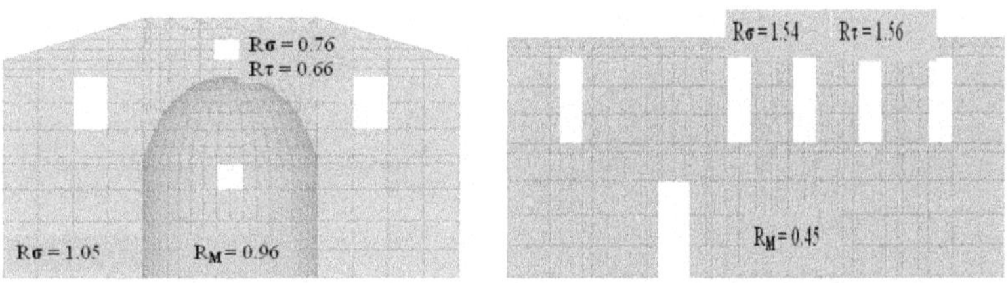

East wall South wall

Figure 9: “Typical Post Byzantine Basilica” Ratio values of strength to demands.

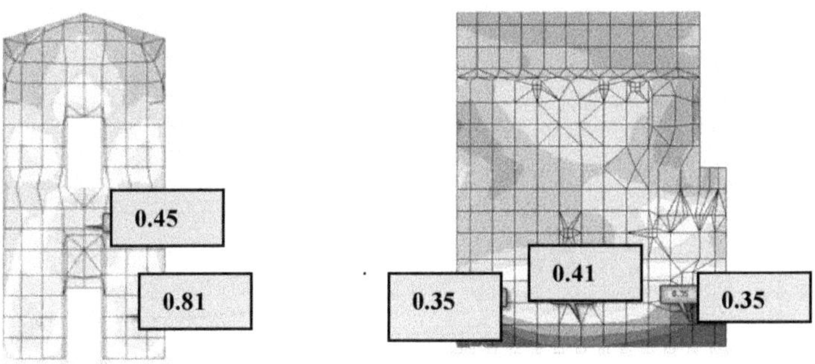

Shear stress Capacity / demand R_τ Non-Isolated $0.9G + 1.4E_y$

Normal stress Capacity / demand Rσ Non-Isolated $0.9G + 1.4E_x$

Figure 10: The church of “The Virgin Mary of Tourniki”.

5 Structural performance with a base isolation

A base isolation system was next introduced to the studied structures representing the "Typical Post Byzantine Basilica", as indicated in fig. 11a as well as to the church of "The Virgin Mary of Tourniki" as indicated in fig. 11b. In fig. 12 the design spectrum considered for the non-isolated (fixed) and the isolated structural systems is shown together with the corresponding fundamental eigen-periods of the non-isolated (fixed) and the isolated structural systems. As can be seen in this figure, the dynamic amplification for the non-isolated and the isolated structures is approximately of the same order. This is due to the response modification factor taking the value $q=1$ for seismic isolated structures instead of $q=1.5$ that was used for un-reinforced masonry structures. The use of more flexible base isolation schemes than the one studied here is a design alternative, which must be further investigated.

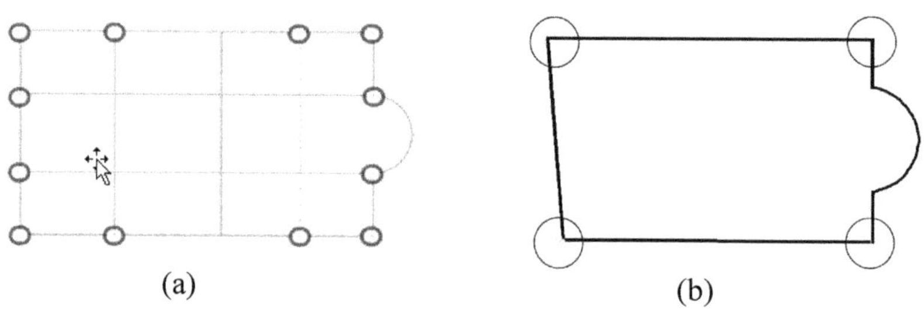

Figure 11: Arrangement of the base isolation scheme. (a) The "Typical Basilica"; (b) "The Virgin Mary of Tourniki".

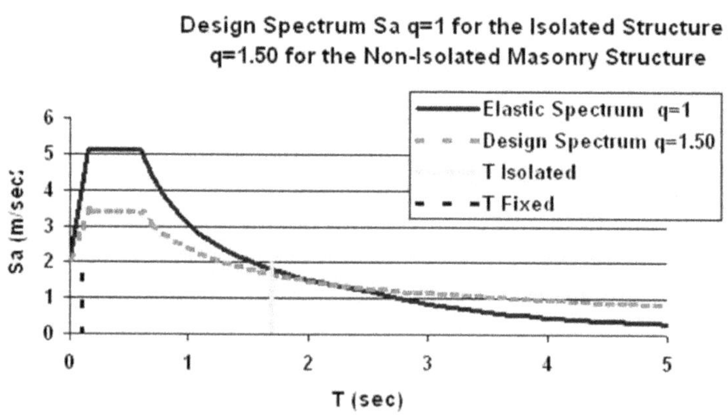

Figure 12: Design spectrum used in the numerical simulations.

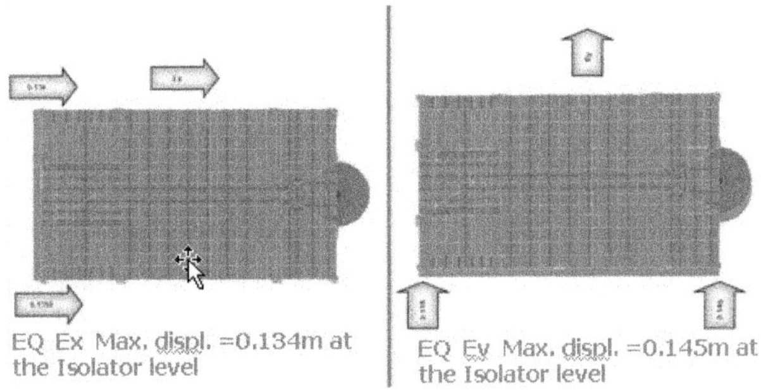

Figure 13: Horizontal displacements of the base isolated "Typical Post Byzantine Basilica".

Figs. 13 and 14 show the deformation demands at the isolation level for the "Typical Post Byzantine Basilica" and the "Virgin Mary of Tourniki", respectively. As can be seen, their maxima are 145 mm for the first structure and 133 mm for the second. Apart from the deformation demands of the base isolation devices, of utmost interest is the effect the introduced base isolation scheme had on the values of the capacity/demand ratio, which were used in section 4 above to predict numerically the observed earthquake structural damage for these churches. In the following section, these ratio values for the non-isolated structures are compared with the corresponding values of the isolated structures.

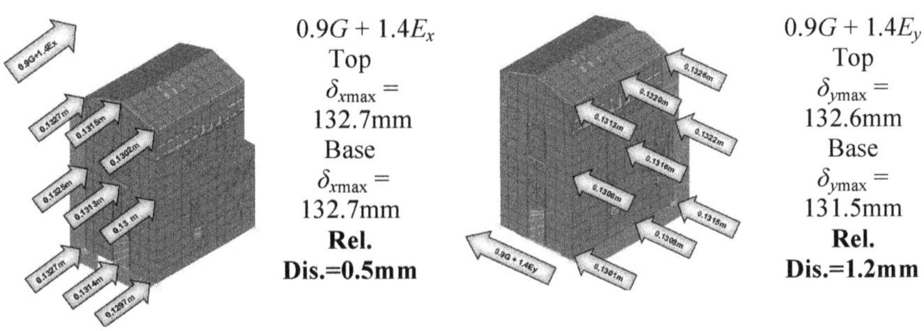

Figure 14: Horizontal displacements of base isolated "Virgin Mary of Tourniki". (4 cylindrical isolators Φ500, Elastomeric height = 148 mm, K_x=K_y=4×1194 kN/m, K_z=4×384 MN/m, K_{zz}=38.4 GN/m, T_{iso}=1.69 s).

5.1 The predicted capacity/demand performance of the isolated structures

The results obtained for the isolated structures are processed in the same way as presented in section 4 by applying the same limit-state criteria that were employed for the non-isolated structures. The results of this evaluation must be viewed with the consideration that there is no reduction in the earthquake loads of the isolated structures, due to the moderate flexibility of the used isolation schemes and the employed design spectrum of fig. 12 combined with the response modification factor value q=1.0 for the isolated structure, as already mentioned. Such a reduction of the earthquake loads will result from the use of a more flexible base isolation scheme than the one employed here.

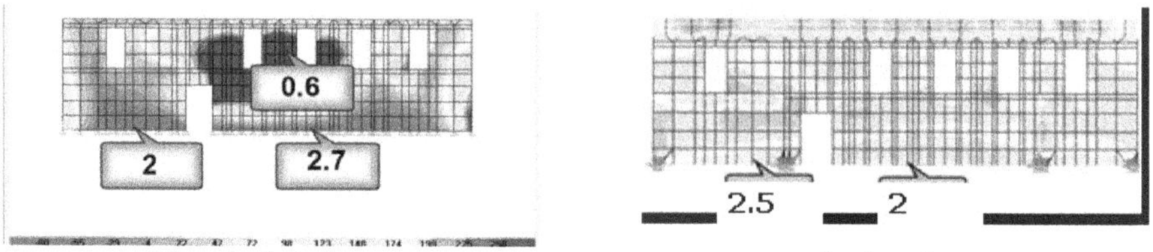

Figure 15: South wall (outside face) $0.9G + 1.4E_y$, out-of-plane S22 strength /demand ratio, "Typical Post Byzantine Basilica".

5.1.1 "Typical Post Byzantine Basilica"

In fig. 15 the strength over demand ratio values for the "Typical Post Byzantine Basilica" structure are shown for both the non-isolated as well as the isolated case. These selected results for the non-isolated structure correspond to ratio values much smaller than 1, which, as explained in section 4, indicates structural damage. As can be seen, for the case of the isolated structure the corresponding strength over demand ratio values at the same locations are well above 1, which indicates that with the introduction of the base isolation scheme structural damage from earthquake loads should be minimized.

5.1.2 "The Virgin Mary of Tourniki"

In fig. 16, the strength to demand ratio values for the "Virgin Mary of Tourniki" are shown for both the non-isolated and the isolated case. The selected results for the non-isolated structure correspond to ratio values much smaller than 1, which represents structural damage. As was the case before, the corresponding strength over demand ratio values at the same locations are well above 1 in almost all cases for the isolated structure. Thus, as expected, structural damage from earthquake loads should be minimized with a base isolation scheme. This improvement in the expected earthquake performance of the studied structural systems will become even better if a more flexible isolation scheme than the one employed here is used. Obviously, the increased flexibility of the base isolation scheme will result in larger deformation demands of the base isolation devices than the ones predicted here (figs. 13 and 14); the selected isolation devices must then pass such relevant checks.

At this point it is important to stress the difficulties in applying a base isolation scheme to an existing structure especially when that structure is not-reinforced masonry, as is the case with both systems investigated here. In order to place the base isolation devices under the existing structure, an additional foundation system must be constructed under the masonry walls that will enable such an operation. Such systems have been tried in the past; however, apart from the difficulty in constructing this additional foundation, the extra cost should also

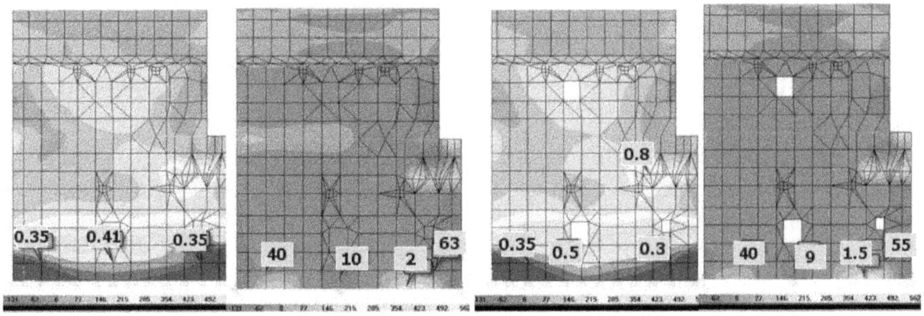

Non-isolated | Isolated | Non-isolated | Isolated

Normal stress strength/demand ratio R_σ , loading $G+E_y$

Figure 16: "The Virgin Mary of Tourniki".

be considered. Another difficulty may arise from older remains frequently found below existing churches, unveiled by the excavation for constructing the additional foundation. This is not the case for the structures investigated here. The church of "The Virgin Mary of Tourniki" has been moved to a different location uphill (fig. 17) due to the flooding of the location, where it was constructed almost 800 years ago, due to the construction of a dam and the filling of its reservoir. In this case, an additional foundation for lifting and sliding this church has being constructed. In this particular case, therefore, the placement of a base isolation scheme under the structure at its new location uphill would be a feasible solution with no significant extra cost and could result in a favourable earthquake performance. Obviously, the base isolation scheme should be considered in addition to all other traditional repair and strengthening measures that are also taking place for this church (fig. 18).

Figure 17: The church on its sliding tracks (end of August 2011).

Figure 18: Interior of the 1st story; masonry consolidated by injections and steel rods prior to the sliding.

6 Conclusions

An attempt is made here to numerically explain the observed seismic behavior of post-Byzantine stone masonry churches. The relatively simplified linear analysis approach together with the adopted failure criteria seems to verify the development of the actual damage in the examined structural formations. Both predicted and actual damage is mainly concentrated at the keys and supports of the arches and vaults, at the supports of the roofing system as well as at the door and window openings and the bases of the peripheral walls. Further extensive verification is needed for the assumed strength values used by the failure criteria.

The numerical introduction of a base-isolation system for the third structural form was quite effective in lowering the demands and thus resulting for most structural elements in Capacity/Demand ratio values larger than one, which is assumed to indicate acceptable structural performance. However, the introduction of base isolation under an existing old stone-masonry structure also faces considerable difficulties, despite the advantages shown by the results of the numerical investigation.

In the particular case of the church of "The Virgin Mary of Tourniki" these difficulties have already been addressed because this old church has been moved uphill to a different location due to flooding from the recent construction of a dam. The first author of this paper has proposed the placement of a base isolation scheme under this structure at its new location. The present study demonstrates that this is a feasible solution in this particular case with no significant extra cost resulting in a favourable earthquake performance, which would be especially beneficial to the precious interior frescos.

References

[1] Manos G. C., Consequences on the urban environment in Greece related to the recent intense earthquake activity. *International Journal of Civil Engineering and Architecture*, **5(12)**, pp. 1065–1090, 2011.

[2] Hatzfeld, D., Karakostas, P., Ziazia, M., Selvaggi, G., Berge, C. & Makropoulos, K., The Kozani-Grevena (Greece) earthquake of May, 13th 1995, a seismological study. *Journal of Geodynamics*, **26(2-4)**, pp. 245-254, 1998.

[3] Theodoulidis, N. & Lekidis, V., The Kozani-Grevena, Northern Greece, earthquake of May 13, 1995: strong motion data and structural response. *Journal of European Earthquake Engineering*, **1**, pp. 3-13, 1996.

[4] Lagomarsino, S., Damage assessment of churches after L'Aquila earthquake (2009). *Bulletin of Earthquake Engineering*, **10(**1), pp. 73-92, 2012.

[5] Modena, C., Casarin, F., Porto, F. & Munari, M., L'Aquila 6th April 2009 earthquake: emergency and post-emergency activities on cultural heritage buildings. *Earthquake Engineering in Europe, vol.* 17, eds. M. Garevski & A. Ansal, Springer: Berlin, pp. 495-521, 2010.

[6] Manos, G.C., Soulis, V. & Diagouma, A., Numerical investigation of the behavior of the church of Agia Triada, Drakotrypa, Greece. *Journal in Advances in Engineering Software*, **39**, pp. 284-300, 2008.

[7] Manos, G.C., Soulis, V., Felekidou, O. & Matsou, V., A numerical investigation of the dynamic and earthquake behavior of Byzantine and post-Byzantine basilicas. *9th U.S. National and 10th Canadian Earthquake Engineering Conference*, Toronto, Canada, 2010.

[8] Manos, G.C. & Karamitsios, N., Earthquake numerical simulation of the dynamic and earthquake behaviour of Greek post-Byzantine churches with and without base isolation. *STREMAH XII*, eds. C.A. Brebbia & L. Binda, WIT Press: Southampton, pp. 709-721, 2011.

[9] Provisions of Greek Seismic Code 2000, OASP, Athens, 1999.

[10] European Committee for Standardization, Euro code 6; Design of Masonry Structures, Part 1-1: General Rules for Building. Rules for Reinforced and Un-reinforced Masonry, EN 1996-1-1:2005.

[11] Chiostrini, S., Galano, L. & Vignoli, A., In-situ shear and compression tests in ancient stone masonry walls in Tuscany, Italy. *Journal of Testing and Evaluation,* **31**(4), pp. 289-303, 2003.

Author Index

Seismic Prevention of Damage

A Case Study in a Mediterranean City

Edited by: ***M. MAUGERI****, Università di Catania, Italy*

The destructive earthquakes of Kocaeli and Düzce in Turkey exemplify the high seismic risk of the Mediterranean area, close to the contact between various tectonic plates and therefore frequently shaken by earthquakes of medium or high intensity.

The scientific, social and political awareness of the seismic risk prompted the Italian Department of Civil Protection to fund a Research Project on earthquake damage scenarios for a high-risk area in the Mediterranean. A subsequent Research Project, focused on detailed scenarios and actions for seismic prevention of damage, was financed by the 'Istituto Nazionale di Geofisica e Vulcanologia' (INGV).

A multidisciplinary team researched the analysis and mitigation of seismic risk in Catania, a mid-sized Mediterranean city subjected to medium-high earthquake risk. This book summarises the results of the two-year project, which involved geology, geophysics, geotechnical, structural and transportation engineering and urban planning.

The book includes material that will be valuable to many specialists including environmental, civil, earthquake and civil defence engineers, geologists, geophysicists and town planners, and will be a useful guideline for seismic hazard assessment and seismic risk evaluation, prevention and mitigation.

Series: Advances in Earthquake Engineering, Vol 14
ISBN: 978-1-84564-004-0 ISBN-10: 1-84564-004-7
Published 2005 / 424pp / £171.00

www.ingramcontent.com/pod-product-compliance
Ingram Content Group UK Ltd.
Pitfield, Milton Keynes, MK11 3LW, UK
UKHW020818290926
15470UKWH00003B/1239

9 781845 647544